NELSON
VICmaths

VCE UNITS ③ + ④

general
mathematics 12

mastery workbook

Dirk Strasser
Greg Neal

Nelson VICmaths General Mathematics 12 Mastery Workbook
1st Edition
Dirk Strasser
Greg Neal
ISBN 9780170464055

Publisher: Dirk Strasser
Additional content created by: ansrsource
Project editor: Alan Stewart
Series cover design: Leigh Ashforth (Watershed Art & Design)
Series text design: Rina Gargano (Alba Design)
Series designer: Nikita Bansal
Production controller: Karen Young
Typeset by: MPS Limited

Any URLs contained in this publication were checked for currency during the production process. Note, however, that the publisher cannot vouch for the ongoing currency of URLs.

Acknowledgements
Selected VCE Examination questions and extracts from the VCE Study Designs are copyright Victorian Curriculum and Assessment Authority (VCAA), reproduced by permission. VCE® is a registered trademark of the VCAA. The VCAA does not endorse this product and makes no warranties regarding the correctness or accuracy of this study resource. To the extent permitted by law, the VCAA excludes all liability for any loss or damage suffered or incurred as a result of accessing, using or relying on the content. Current VCE Study Designs, past VCE exams and related content can be accessed directly at www.vcaa.vic.edu.au.

TI-Nspire: Images used with permission by Texas Instruments, Inc
Casio ClassPad: Shriro Australia Pty. Ltd.

For product information and technology assistance,
in Australia call **1300 790 853**;
in New Zealand call **0800 449 725**

For permission to use material from this text or product, please email **aust.permissions@cengage.com**

ISBN 978 0 17 046405 5

Cengage Learning Australia
Level 7, 80 Dorcas Street
South Melbourne, Victoria Australia 3205

Cengage Learning New Zealand
Unit 4B Rosedale Office Park
331 Rosedale Road, Albany, North Shore 0632, NZ

For learning solutions, visit **cengage.com.au**

Printed in China by 1010 Printing International Limited.
1 2 3 4 5 6 7 26 25 24 23 22

Contents

Loans, investments and finance solvers   **89**

To the student

Nelson VICmaths is your best friend when it comes to studying General Mathematics in Year 12. It has been written to help you maximise your learning and success this year. Every explanation, every exam hack and every worked example has been written with the exams in mind.

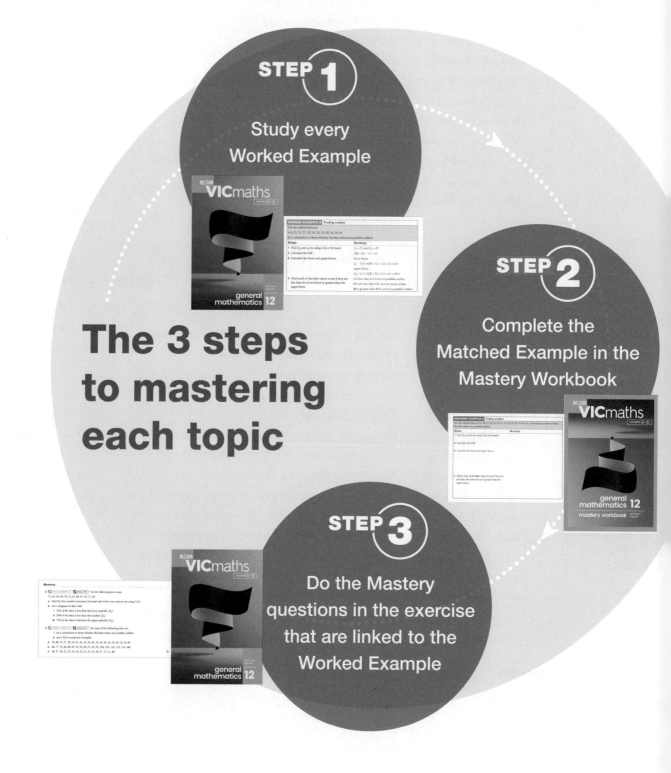

The 3 steps to mastering each topic

STEP 1
Study every Worked Example

STEP 2
Complete the Matched Example in the Mastery Workbook

STEP 3
Do the Mastery questions in the exercise that are linked to the Worked Example

CHAPTER 1

DATA DISTRIBUTIONS

SB

p. 4

MATCHED EXAMPLE 1	Classifying variables

State whether the following variables shown in *italics* are
   i numerical or categorical
   ii discrete, continuous, ordinal or nominal	

Steps	**Working**
a *Taste* of food in a restaurant (good, average, bad)	
i Are there numbers involved?	
ii Is there a natural order?	
b *Number* of children in a family	
i **1** Are there numbers involved?	
2 Does it make sense to add the numbers?	
ii Can it be measured with increasing levels of accuracy?	
c *Age* of a teacher (less than 30 years, 30 years, greater than 30 years)	
i **1** Are there numbers involved?	
2 Does it make sense to add the numbers?	
ii Is there a natural order?	
d *Height* of a student	
i **1** Are there numbers involved?	
2 Does it make sense to add the numbers?	
ii Can it be measured with increasing levels of accuracy?	
e *Name* of a school	
i Are there numbers involved?	
Is there a natural order?	

MATCHED EXAMPLE 2 | Displaying categorical variables

The following is the raw data of the types of juices the first 30 people ordered one day at the Auburn Juice Lounge, where A = Apple, S = Strawberry, O = Orange, W = Watermelon, C = Carrot and P = Pineapple.

S, C, A, W, O, O, W, P, S, S, W, W, A, S, O, S, P, S, A, S, W, S, C, S, A, P, W, C, O, A

Steps	Working

a Set up a frequency table that includes both the frequency and the percentage of each type of juice ordered.

1 Set up a table with three columns and list the categories in the first column. Count the number in each category and record the frequency.

Check that the total frequency equals the total number of data values given in the question.

Juice	Frequency	Percentage
Apple		
Strawberry		
Orange		
Watermelon		
Carrot		
Pineapple		
Total		

2 Calculate the percentage for each category using

$$\text{percentage} = \frac{\text{frequency}}{\text{total}} \times 100\%$$

Check that the total percentage is equal to 100% (or 99.9% or 100.1% if the percentages have been rounded).

Answer in the above table.

b Draw a bar chart of this frequency table showing the number of each type of juice ordered, with the categories on the horizontal axis.

Draw a bar chart with the categories on the horizontal axis.

Add a title, label the horizontal axis with the variable name, and label the vertical axis 'Frequency'.

Note: As the variables are categorical nominal and therefore don't have a natural order, the bars can appear in any order.

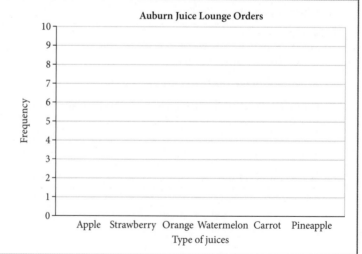

c Draw a percentaged segmented bar chart of this frequency table.

Draw a percentaged segmented bar chart with the vertical axis labelled 'Percentage'.

Add a title and a key matching the order in which the categories appear in the bar chart.

d State the most frequently occurring category.

Find the category that occurs most often.

p. 8

MATCHED EXAMPLE 3 | Finding measures of centre and spread

For each of the following data sets, find the

| i mode | ii mean | iii median | iv range. |

| **Steps** | **Working** |

a The number of bulbs used in seven households:

1, 3, 5, 9, 3, 9, 7, 6

  i Find the most frequently occurring data value(s).

  ii Divide the sum of all the data values by the total number of values.

  iii **1** Order the n data values from smallest to largest.

    2 Find n and note if the number is odd or even.

    3 Find the position of the median.

    4 If n is odd, find the middle value.

      If n is even, average the two middle values.

  iv Range = largest value − smallest value

b The maximum daily temperatures (°C to one decimal place) in a city during a week in August:

4.1, 2.7, 0.0, −1.3, −4.2, 2.7, 1.8

  i Find the most frequently occurring data value(s).

  ii Dividing the sum of all the data values by the total number of values.

  iii **1** Order the n data values from smallest to largest.

    2 Find n and note if the number is odd or even.

    3 Find the position of the median.

    4 If n is odd, find the middle value.

      If n is even, average the two middle values.

  iv Range = largest value − smallest value

SB
p. 16

The following histogram shows the heights of people (in cm) in a small town.

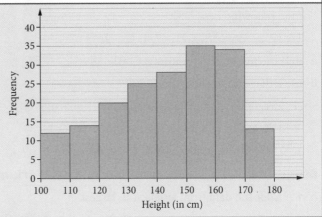

Steps	Working
a How many people's heights range between 120 and 140 cm?	
Find the frequencies from the histogram.	
b Is the histogram approximately symmetric, positively skewed or negatively skewed? Does it have any possible outliers?	
The histogram has a negative tail.	
c What is the modal interval?	
Which height group has the highest frequency?	
d In which interval is the median?	

1 Create a grouped frequency table for the histogram.

Height	Frequency
100–<110	.
110–<120	
Total	

2 Count the number of data values n. Note whether it is odd or even, and find the position of the median.

3 Add the frequencies in order from the grouped frequency table until the interval with the median position is reached.

SB

Using CAS 1:
Constructing
a histogram
from raw data
p. 17

MATCHED EXAMPLE 5 | Finding the five-number summary by hand

For the following data

    12, 17, 21, 56, 13, 14, 21, 47, 4, 22, 29, 33

a find the five-number summary by hand

b use a diagram to show that

   i 25% of the data is less than the lower quartile (Q_1).

   ii 50% of the data is less than the median (Q_2).

   iii 75% of the data is less than the upper quartile (Q_3).

Steps	Working
a 1 Order the data from smallest to largest.	
2 Find the minimum and maximum value.	
3 Find the median. There is an even number of data values, so find the average of the two middle points.	
4 Find Q_1, the median of the lower half of the data.	
5 Find Q_3, the median of the upper half of the data.	
6 List the five-number summary.	
b 1 Draw a diagram showing the three quartiles.	
2 Use the diagram to calculate the percentage of data less than each quartile.	**i**
	ii
	iii

SB

Using CAS 2:
Finding the five-
number summary
p. 26

| **MATCHED EXAMPLE 6** | Finding outliers |

For the ordered data set 13, 16, 17, 18, 20, 20, 21, 23, 24, 26, 26, 35, 45, do a calculation to show whether the blue values are possible outliers.

Steps	**Working**
1 Find Q_1 and Q_3 by using CAS or by hand.	
2 Calculate the IQR.	
3 Calculate the lower and upper fences.	
4 Check each of the blue values to see if they are less than the lower fence or greater than the upper fence.	

SB

Using CAS 3:
Constructing
boxplots p. 28

MATCHED EXAMPLE 7 | Reading boxplots

The boxplot shows the distribution of 60 student test scores marked out of 30.

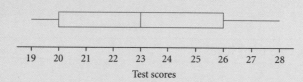

Find the

a five-number summary

b percentage of employees who scored more than 26

c percentage of employees who scored less than 23

d percentage of employees who scored between 19 and 26

e number of employees who scored less than 20

f scores at the lower end that would be considered outliers

g scores at the upper end that would be considered outliers.

Steps	Working
a Read directly from the boxplot.	
b–d Use the fact that quartiles divide data into four equal groups, so 25% of the data is in each group. 25% · 25% · 25% · 25%	
e Find the percentage first and then multiply by the total number.	
f Use the IQR to calculate the lower fence. lower fence $= Q_1 - 1.5 \times IQR$	
g Use the IQR to calculate the upper fence. upper fence $= Q_3 + 1.5 \times IQR$	

MATCHED EXAMPLE 8 Rounding to decimal places versus significant figures

Round each number to

   i two decimal places   **ii** two significant figures

a 7.111   **b** 4.429   **c** 0.7847   **d** 28 847

e 86.664   **f** 42 478.077   **g** 57.7878

Steps	Working
a **i** Focus on the first two decimal places.	
ii Focus on the first two significant figures.	
b **i** Focus on the first two decimal places.	
ii Focus on the first two significant figures.	
c **i** Focus on the first two decimal places.	
ii Focus on the first two significant figures.	
d **i** Focus on the first two decimal places.	
ii Focus on the first two significant figures.	
e **i** Focus on the first two decimal places.	
ii Focus on the first two significant figures.	
f **i** Focus on the first two decimal places.	
ii Focus on the first two significant figures.	
g **i** Focus on the first two decimal places.	
ii Focus on the first two significant figures.	

MATCHED EXAMPLE 9 | Reading histograms with log scales

The histogram shows the number of views on 77 YouTube videos of an independent YouTuber during one week.

a How many videos had

   i over 100 000 views during the week?

   ii over 10 000 views during the week?

   iii under 1000 views during the week?

b What percentage of videos had between 100 and 1000 views during the week? Round to three significant figures.

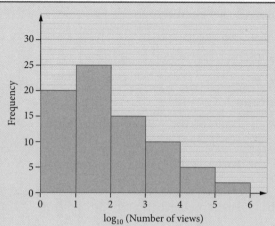

Steps	Working
1 a Rewrite the log scale to show the actual values.	
2 Read the result using these actual values.	**i** **ii** **iii**
b Read the result using these actual values and convert it to a percentage, using $$\text{percentage} = \frac{\text{frequency}}{\text{total}} \times 100\% \text{ and}$$ rounding to three significant figures.	

9780170464055

MATCHED EXAMPLE 10 | Using dot plots

SB

p. 42

The dot plot shows the marks of students scored in an examination.

a Find the

    i mode   **ii** range

    iii median   **iv** lower quartile (Q_1)

    v upper quartile (Q_3)   **vi** interquartile range (IQR)

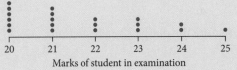

Marks of student in examination

b What could best describe the shape of the distribution: approximately symmetric, positively skewed, or negatively skewed?

Steps	Working
a **i** Find the most common value.	
ii range = largest value − smallest value	
iii **1** Count the number of dots n, note whether n is odd or even, and find the position of the median.	
2 If n is odd, find the data value of the middle dot. If n is even, find the average of the data values for the two middle dots. Count each column of dots from the bottom up to reach the median.	
3 Write the answer.	
iv **1** To find the lower quartile Q_1, find the median of the lower half. Count the number of dots n, note whether n is odd or even, and find the position of Q_1.	
2 If n is odd, find the data value of the middle dot. If n is even, find the average of the data values for the two middle dots. Count each column of dots from the bottom up to reach the lower quartile.	
3 Write the answer.	
v **1** To find the upper quartile Q_3, find the median of the upper half. Count the number of dots n, note whether n is odd or even, and find the position of Q_3.	
2 If n is odd, find the data value of the middle dot. If n is even, find the average of the data values for the two middle dots. Count each column of dots from the bottom up to reach the upper quartile.	

3 Write the answer.

vi Use Q_3 and Q_1 to calculate the interquartile range.

b Picture the dot plot as a histogram.

9780170464055

MATCHED EXAMPLE 11 | Using stem plots

The stem plot shows the runs scored by a player in the last 30 cricket matches.

a Find the

i mode	**ii** range
iii median	**iv** lower quartile (Q_1)
v upper quartile (Q_3)	**vi** interquartile range (IQR)

b Is there an outlier? Justify your answer.

Stem	Leaf
0	2 5 6
1	4 6 8 9 9
2	1 1 2 4 4 8 8
3	0 3 4 5 5 6
4	0 2 3 3 3 4 5 7 8

p. 44

Steps	Working
a i Find the most common value.	
ii range = largest value − smallest value	
iii 1 Count the number of data values *n*, note whether it's odd or even, and find the position of the median.	
2 If *n* is odd, find the middle data value. If *n* is even, find the average of the two middle data values.	
3 Write the answer.	
iv 1 To find the lower quartile Q_1, find the median of the lower half. Count the number of data values *n*, note whether it's odd or even, and find the position of Q_1.	
2 If *n* is odd, find the middle data value. If *n* is even, find the average of the two middle data values.	
3 Write the answer.	

Using CAS 4:
Finding the mean and standard deviation for ungrouped data
p. 52

Using CAS 5:
Finding the mean and standard deviation for grouped data
p. 53

v 1 To find the upper quartile Q_3, find the median of the upper half.

Count the number of data values n, note whether it's odd or even, and find the position of Q_3.

2 If n is odd, find the middle data value.

If n is even, find the average of the two middle data values.

3 Write the answer.

vi Use Q_3 and Q_1 to calculate the interquartile range.

b Check any value that appears to be an outlier against the upper or lower fence

MATCHED EXAMPLE 12 | Working with the mean and standard deviation from a display

SB

p. 54

For each the following displays

　i　find the mean

　ii　find the median

　iii　state whether the mean or median or both are appropriate measures of the centre and why

　iv　find the standard deviation.

　v　find the IQR

　vi　state whether the standard deviation or IQR or both are appropriate measures of the spread and why.

Round your answers to one decimal place where necessary.

Steps	Working
a　Number of toys bought in a store by 17 children	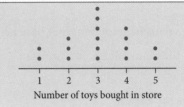
i　Use CAS by entering the values 1, 1, 2, 2, 2, 3, 3 … and selecting the mean.	
ii　Use CAS and select the median.	
iii　Is the distribution approximately symmetric or skewed? Does the distribution have outliers?	
iv　Use CAS and select the standard deviation.	
v　Use CAS to find the IQR.	
vi　Is the distribution approximately symmetric or skewed? Does the distribution have outliers?	
b　Number of cars that passed a toll in each hour in the last 24 hours	Stem | Leaf 1 | 1 2 4 5 6 6 7 8 2 | 1 1 2 2 3 4 5 3 | 0 3 4 4 5 4 | 0 5 8 5 | 6 Key: 1|3 = 13
i　Use CAS by entering the values 11, 12, 14, 15 … and selecting the mean.	
ii　Use CAS and select the median.	
iii　Is the distribution approximately symmetric or skewed? Does the distribution have outliers?	
iv　Use CAS and select the standard deviation.	
v　Use CAS to find the IQR.	
vi　Is the distribution approximately symmetric or skewed? Does the distribution have outliers?	

MATCHED EXAMPLE 13 | Working with the 68–95–99.7% rule

The lifespan of lions in a particular zoo approximated a bell-shaped distribution with a mean of 14 years and a standard deviation of 2 years.

Steps	Working
a Write the 68–95–99.7% rule scale for this.	
Write a 68–95–99.7% rule scale that includes the mean and standard deviations given.	
b Find the percentage of lions that have a lifespan of between 10 and 18 years.	
1 Add the required percentages from the 68–95–99.7% rule scale.	
2 Write the answer.	
c Find the percentage of lions having a lifespan of greater than 16 years.	
1 Add the required percentages from the 68–95–99.7% rule scale.	
2 Write the answer.	
d Find the percentage of lions having a lifespan of greater than 12 years.	
1 Add the required percentages from the 68–95–99.7% rule scale.	
2 Write the answer.	
e Assume that the pairs of lions were bred in the zoo and the numbers after breeding grew up to 3000.	
i How many of these would they expect to have a lifespan of lesser than 8 years?	
1 Add the required percentages from the 68–95–99.7% rule scale.	
2 Find this percentage of the total given and write the answer.	
ii How many of these would they expect to have a lifespan of between 8 and 20 years?	
1 Add the required percentages from the 68–95–99.7% rule scale.	
2 Find this percentage of the total given and write the answer.	

MATCHED EXAMPLE 14 | Working with *z*-scores

The income of a burger sale stand is normally distributed with a mean of $400 and a standard deviation of $100. The income of the burger sale stand on a particular day is $500.

Steps	Working
a Calculate the standardised value for this actual value.	
1 Write the *z*-score formula.	
2 Substitute the actual value, mean and standard deviation.	
b Show that this standardised value means that 84% of the time the sales are less than $500.	
1 State how the *z*-score relates to the number of standard deviations.	
2 Sketch the relevant standardised 68–95–99.7% rule diagram and show the required region or use the 68–95–99.7% rule scale.	
3 Read the percentage from the diagram.	
c If on another day the standardised sales is $z = 2$, what is actual value of the sales?	
1 Write the formula for finding the actual value from the standardised value, and the values for z, s_x and $\overline{x}$.	
2 Substitute the values into the formula and solve for x.	
3 Write the answer.	

MATCHED EXAMPLE 15 | Using *z*-scores to compare

The table below shows the marks out of 100 that a student has achieved on his final exams in three subjects and the means and standard deviations for each of the subjects.

	Mark	Mean	Standard deviation
Physics	78	68	5
Computer science	87	75	4
Chemistry	77	74	3

Assume the results for each of the three subjects approximates a bell-shaped distribution.

Steps	Working

a In which subject did the student perform the best?

  1 Write the *z*-score formula.

  2 Substitute the values for each subject or use the 68–95–99.7% rule scale.

  3 State which *z*-score is the largest.

b In which of these subjects was he in the top 0.15% of students? Draw a diagram that shows how you obtained your answer.

  1 Sketch the relevant standardised 68–95–99.7% rule diagram.

  2 Read the answer from the diagram or use the 68–95–99.7% rule scale.

CHAPTER ②

ASSOCIATIONS BETWEEN TWO VARIABLES

SB

p. 86

MATCHED EXAMPLE 1	Identifying explanatory and response variables

For each of the following

   i identify the two variables

   ii state whether each one is categorical or numerical

   iii identify the explanatory variable, giving a reason for your answer.

Steps	Working

a Studies are undertaken to establish if the temperature of water in a lake explains the number of fish in the lake.

   i Consider the sort of data that will be collected.

   ii Are there numbers involved?

     Does it make sense to add the numbers?

   iii Which variable is 'explaining' or 'predicting' the other? If the wording isn't used, which variable is most likely to affect the other?

---

b A scientist wants to compare the effect of three different hair serums on the hair growth of a person.

   i Consider the sort of data that will be collected.

   ii Are there numbers involved?

     Does it make sense to add the numbers?

   iii Which variable is 'explaining' or 'predicting' the other? If the wording isn't used, which variable is most likely to affect the other?

---

c A study is undertaken to see whether the origin of the tea leaves (China, India, Sri Lanka) affects the tea's stimulating effect (scale of 0 to 5, where 0 is none and 5 is extremely high).

   i Consider the sort of data that will be collected.

   ii Are there numbers involved?

     Does it make sense to add the numbers?

   iii Which variable is 'explaining' or 'predicting' the other? If the wording isn't used, which variable is most likely to affect the other?

d An experiment is conducted to study how different workout intensities (high, moderate, low) affects weight loss.

   i Consider the sort of data that will be collected.

   ii Are there numbers involved?

     Does it make sense to add the numbers?

   iii Which variable is 'explaining' or 'predicting' the other? If the wording isn't used, which variable is most likely to affect the other?

9780170464055

| MATCHED EXAMPLE 2 | Creating two-way frequency tables |

Thirty children and twenty adults were interviewed about their movie preferences. Eleven children preferred watching live-action movies, while five of the adults preferred animated movies.

a Name the explanatory and the response variables.

b Construct a two-way frequency table to show this information using *age group* for the columns.

Steps	Working
a Name the explanatory and the response variables.	
b 1 Create a table using *age group* as the explanatory variable for the columns and *movie preference* as the response variable for the rows.	
2 Fill in the information from the question.	Answer in the table you've drawn in step 1.
3 Complete the table using column and row totals.	Answer in the table you've drawn in step 1.

MATCHED EXAMPLE 3 Working with percentage two-way frequency tables

a Convert the following two-way frequency table into a percentage two-way frequency table by percentaging the explanatory variable. Round to the nearest percentage.

b What does the table suggest about the association between age and movie preference? Give reasons by referring to percentages.

Movie preference	Age group		
	Children	Adult	Total
Animated	32	12	44
Live-action	13	28	41
Total	45	40	85

Steps	Working

a **1** The explanatory variable forms the columns, so redraw the two-way table using only the column totals.

2 Calculate the required percentages.

3 Put the percentages into the two-way table.

Movie preference	Age group	
	Children	Adult
Animated		
Live-action		
Total		

b An association means the explanatory variable categories give considerably different results. Refer to percentages in your answer.

| MATCHED EXAMPLE 4 | Reading parallel percentage segmented bar charts |

The following parallel percentage segmented bar chart shows a survey result of the demand for oranges and grapes in different seasons. Discuss if this suggests there is an association between two different fruits and seasons by comparing percentages.

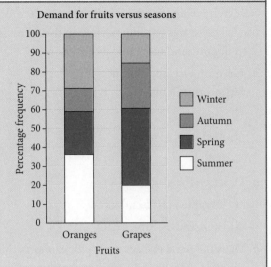

Demand for fruits versus seasons

Legend:
- Winter
- Autumn
- Spring
- Summer

Steps	Working
Look at how many of the segments are similar and how many are different.	

MATCHED EXAMPLE 5 | Interpreting back-to-back stem plots

Research is done to establish if there is a connection between the growth of a tree and the type of fertiliser used. The results are summarised in the following back-to-back stem plot with 22 weeks of tree growth with organic and inorganic fertiliser.

a For both Organic and Inorganic fertilisers, calculate the

   i range

   ii median

   iii IQR.

b Describe the shape of the

   i Organic Fertiliser data

   ii Inorganic Fertiliser data.

c Does the back-to-back stem plot supports the contention that the height of the tree is associated with the type of fertiliser? Refer to the values of three appropriate statistics and the shape of the data in your response.

Organic Fertiliser		Inorganic Fertiliser
5	0	1
6 4 2 2 2 2 0 0	1	2 4
8 8 6 6 2 0 0	2	4 6 6
2 2 1 0	3	2 4 6 8
0 0	4	0 2 2 2 4 6
	5	0 2 2 4 6 9
Key: 0|4 = 4.0 m		Key: 3|2 = 3.2 m

Steps	Working
a **i** range = largest value − smallest value	
ii If the number of data values is odd, find the middle data value. If the number of data values is even, find the average of the two middle data values.	
iii IQR = $Q_3 - Q_1$	
b **i** To see the shape of the right leaf data, rotate the page 90° anticlockwise so that the stem forms the horizontal axis and picture it as a histogram.	

ii To see the shape of the left leaf data, picture the data
on the right and repeat the right leaf data steps.

c Compare the ranges, medians, IQRs and the shapes.
Do the two stem plots show differences?

MATCHED EXAMPLE 6 | Interpreting parallel dot plots

A small chocolate company analysed the lolly sales for the first three quarters of the previous year to analyse the association between lolly sales and the period of sales.

a Describe the shape of each distribution.

b Calculate the median, range and IQR for each quarter.

c Do the dot plots support the contention that the lolly sales of the chocolate company are associated with the period of sales? Quote the IQR values in your response.

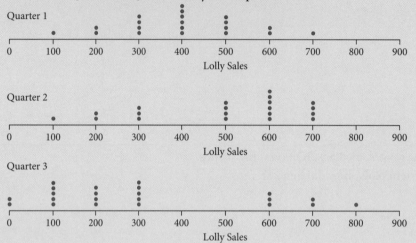

Steps	Working
a Comment on symmetry and skewness for each distribution.	
b Calculate the median, range and IQR from each of the parallel dot plots.	
c Compare the IQRs to see if the dot plots show differences.	

MATCHED EXAMPLE 7 | Interpreting parallel boxplots

A bookstore owner has three different stores in different locations. The owner studies the data recorded in the last 12 months to check the association between the number of books sold and the location of the stores. The results are summarised in the following parallel boxplots.

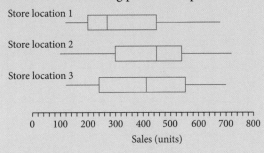

a Describe the shape of each distribution.

b Calculate the median, range and IQR for each store location.

c Do the boxplots support the contention that the number of books sold is associated with the store location? Quote the shape and the range of the data in your response.

Steps	Working
a Comment on symmetry and skewness for each distribution.	
b Calculate the median, range and IQR from each of the parallel boxplots.	
c Compare the shapes and ranges to see if the boxplots show differences.	

MATCHED EXAMPLE 8 | Interpreting scatterplots

A study was conducted on the association between the development of a large breed's puppies (months) and their weight (kg). The figure below shows a scatterplot for the weight of the large breed's puppies versus their age.

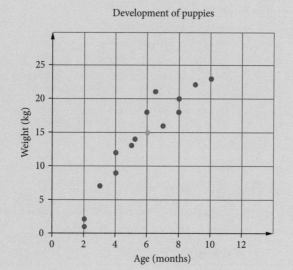

Development of puppies

Steps	Working
a What is the explanatory variable?	
The explanatory variable appears on the *x*-axis.	
b What is the response variable?	
The response variable appears on the *y*-axis.	
c How many puppies were there in the study?	
Count the number of dots.	
d What does the pale blue dot represent?	
Read from both axes.	
e In the study, how many puppies were less than 6 months old?	
Read from the *x*-axis.	
f How old is the heaviest puppy?	
Read from the *y*-axis.	

MATCHED EXAMPLE 9 | Describing association using scatterplots

For each of the following scatterplots

  i describe the association between the two variables in terms of direction, form and strength

  ii explain what this means in terms of the variables.

a

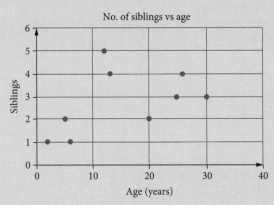

No. of siblings vs age

b

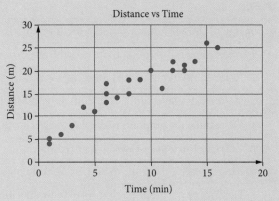

Distance vs Time

c

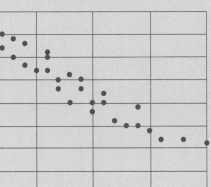

Jumpers sold vs temperature

Steps	Working
a **i** Is the data sloping up or down? Does the data follow a straight line pattern? How spread out are the data points? **ii** Refer to the variables.	
b **i** Is the data sloping up or down? Does the data follow a straight line pattern? How spread out are the data points? **ii** Refer to the variables.	
c **i** Is the data sloping up or down? Does the data follow a straight line pattern? How spread out are the data points? **ii** Refer to the variables.	

SB

Using CAS 2:
Constructing
scatterplots
p. 113

MATCHED EXAMPLE 10 | Choosing a graph for two variables

For the following

    i determine which graph(s) can be used to display the association between the two variables

    ii give a reason for your answer.

Steps	Working
a the age of the cricket players in a team and their batting average	
i Determine the graph(s) to display the association	
ii Refer to the types of variables involved.	
b eye colours of players in a cricket match and their home states	
i Determine the graph(s) to display the association	
ii Refer to the types of variables involved.	
c whether a person is from Sydney or Melbourne and the number of books they read in a year	
i Determine the graph(s) to display the association	
ii Refer to the types of variables involved.	
d different schools and math test scores of grade 9.	
i Determine the graph(s) to display the association	
ii Refer to the types of variables involved.	

MATCHED EXAMPLE 11 | Interpreting correlation coefficient values

Interpret the correlation coefficient values and write a sentence describing the association for each of the studies described below, beginning with 'The data suggests'.

Steps	Working
a A study is investigating whether there is an association between a person's *age* and *the number of pets* they own. The correlation coefficient was found to be $r = 0.152$.	
What level of strength of association does r indicate? Is the association positive or negative?	
b A study is investigating whether there is an association between the *number of coffees* a person drinks and the *number of hours* they sleep. The correlation coefficient was found to be $r = -0.712$.	
What level of strength of association does r indicate? Is the association positive or negative?	
c A study is investigating whether there is an association between the *length* of the train and the *time* it takes to cross a railway station. The correlation coefficient was found to be $r = 0.899$.	
What level of strength of association does r indicate? Is the association positive or negative?	

SB

Using CAS 3:
Calculating
the Pearson
correlation
coefficient
p. 120

MATCHED EXAMPLE 12 | Exploring causation

For each of the following correlations between pairs of variables, find another variable that could be the underlying cause of the correlation between the two.

Steps	Working
a A positive correlation between *the number of pastas sold in a restaurant* and *the number of cars parked in the parking lot of a restaurant*	
Which variable might be causing changes in both?	
b A positive correlation between *the average acreage burned annually by bush fires* and *the death rates of emperor penguins*	
Which variable might be causing changes in both?	
c A positive correlation between *the number of hours a laptop is charged* and *the number of hours spent working on the laptop*	
Which variable might be causing changes in both?	
d A positive correlation between *the foot length of a person* and *the weight of a person*	
Which variable might be causing changes in both?	

9780170464055

LINEAR ASSOCIATIONS

MATCHED EXAMPLE 1	Finding the least squares line of best fit equation using the formula

The total runs scored and the total balls faced by a batsman over a cricket season were recorded and the values of the following statistics were determined:

Sample mean of the runs scored was 10.02

Sample standard deviation of the runs scored was 9.45

Sample mean of the balls faced was 20.88

Sample standard deviation of the balls faced was 16.21

The Pearson correlation coefficient was 0.56

a State the explanatory and response variables.

b Write the values for $\bar{x}$, s_x, $\bar{y}$, s_y and r.

c Calculate the equation for the least squares line of best fit that models this data, rounding the coefficients a and b to two significant figures.

SB
p. 143

Steps	Working
a Which variable affects the other?	
b Which is the x-variable and which is the y-variable?	
c 1 Calculate the slope (b) using the formula and round to two significant figures.	
2 Calculate the y-intercept (a) using the formula and round to two significant figures. Use the *unrounded* answer for b when calculating the value of a.	
3 Write the equation for the least squares line of best fit.	
4 Replace y and x in the equation with the correct variable names.	

SB

Using CAS 1:
Finding the least squares line of best fit equation
p. 144

SB

Using CAS 2:
Graphing the least squares line of best fit
p. 145

MATCHED EXAMPLE 2 | Finding the least squares line of best fit

Peter is conducting an experiment by heating a flask of water. The temperature of the water is recorded at one-minute intervals. The resulting least squares line of best fit is fitted to a scatterplot. Find

a the y-intercept and interpret what it means

b the slope rounded to three significant figures and interpret what it means

c the least squares line of best fit equation.

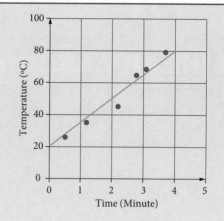

Steps	Working
a 1 Where does the line of best fit cross the y-axis? **2** Interpret what the y-intercept means in terms of the variables and units.	
b 1 Choose two easy-to-identify points on the line. **2** Calculate the slope from these two points and give the answer to the required rounding. **3** Interpret what the slope means in terms of the variables and units.	
c The least squares line of best fit equation is $\quad y = a + bx,$ where a is y-intercept and b is the slope. Use the variables given instead of x and y.	

MATCHED EXAMPLE 3 Interpreting the least squares line of best fit

SB

p. 147

For the following least squares line of best fit equation

$$productivity = -5.61 + 2.73 \times man\text{-}hours$$

where productivity is measured in the number of units and man-hours is measured in hours,

a identify and interpret the slope

b identify and interpret the y-intercept and comment on your result.

Steps	Working
a The slope is the value that man-hours is multiplied by in the equation. Interpret the slope in terms of the variables and units.	
b Identify and interpret the y-intercept and comment.	

SB

Using CAS 3:
Finding the
coefficient of
determination
p. 153

SB

p. 154

MATCHED EXAMPLE 4	Interpreting the coefficient of determination

Data was collected to investigate the association between the number of motorcycles and the number of cars that crossed a bridge in five-minutes intervals and is displayed in the table below.

Number of motorcycles	71	54	70	59	51	48	73	66
Number of cars	80	67	79	71	65	58	90	87

a Assuming that the *number of motorcycles* is the explanatory variable, calculate the coefficient of determination, correct to three decimal places.

b Interpret the coefficient of determination.

c What percentage of variation in the number of cars is not explained by the variation in the number of motorcycles? Round your answer to the nearest whole number.

d What is the least squares line of best fit equation that models this data? Round the coefficients a and b to two significant figures.

e Do you think that the model is appropriate? Justify your answer.

Steps	**Working**
a We are told that the number of cars is the explanatory variable. If we weren't told, we would have had to determine this. Use a CAS to find r^2. The a and b values for the least squares line of best fit equation will also be calculated.	**TI-Nspire**　　　　　**ClassPad**
b Calculate $r^2 \times 100\%$, then interpret the result using the general sentence.	
c Subtract the coefficient of determination percentage from 100% and round to the nearest whole number.	
d 1 Read the a and b values for the line of best fit equation from the screen and write the equation using these values.　　**2** Round a and b to two significant figures.　　**3** Replace x and y in the equation with the correct variable names.	
e Determine the appropriateness of the model using r^2 to support your decision.	

MATCHED EXAMPLE 5 | Finding the correlation coefficient from the coefficient of determination

Find the value of the Pearson correlation coefficient, correct to two decimal places, for each of the following:

Steps	Working
a The coefficient of determination for the data displayed in the scatterplot is 0.81.	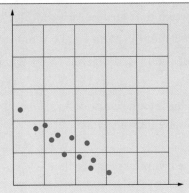

1 Calculate the square root of the coefficient of determination rounded to the given decimal places. Remember that this value could be positive or negative.

2 Is the association shown on the scatterplot positive or negative?

3 The association allows us to determine if r is positive or negative.

b The least squares line of best fit that enables the goals scored in the football match and the number of matches played is

number of goals = 0.2 + 1.3 × number of matches

The coefficient of determination for this data is 0.72.

1 Calculate the square root of the coefficient of determination rounded to the given decimal places. Remember that this value could be positive or negative.

2 Find the slope of the least squares line of best fit and check if it is positive or negative.

3 The sign of the slope allows us to determine if r is positive or negative.

MATCHED EXAMPLE 6 | Making predictions from the line of best

Data was collected from children aged between 1 and 10 years and a least squares line of best fit was found to have the equation

$$TV\ watch\ time\ \text{(minutes)} = 12.5 + 35 \times age\ \text{(years)}$$

a Predict the TV watch time for a 2-year-old. Does this involve interpolation or extrapolation?

b Predict the TV watch time for a 12-year-old. Does this involve interpolation or extrapolation?

c Which of the predictions in parts **a** and **b** is more reliable? Justify your answer.

d Predict the age of a child who watches TV for 310 minutes.

Steps	**Working**
a Substitute the value into the equation in place of *age* and solve for *watch time*. Was the value used within or outside the original data range?	
b Substitute the value into the equation in place of *age* and solve for *watching time*. Was the value used within or outside the original data range?	
c Decide which of the two is the more reliable prediction and justify your decision.	
d Substitute the value into the equation in place of *TV watch time* and solve for *age*, using CAS if necessary.	

TI-Nspire

1 Press **menu** > **Algebra** > **Solve**.

2 Enter the equation followed by, **x**, where *x* represents the variable *age*.

3 Press **enter**.

ClassPad

1 Tap **Interactive** > **Advanced** > **solve**.

2 In the dialogue box, enter the equation in the **Equation:** field.

3 In the **Variable field**, keep the default variable **x**.

4 Tap **OK**.

MATCHED EXAMPLE 7 | Finding residual values from the line of best fit

SB
p. 168

Find the residual values for each of the following.

Steps	Working

a For a study done to establish the association between the salary of a person ($) and the average spent ($) in a month, find the residual values for the following directly from the graph:
 	
  i Robert, whose salary is $2000.
 	
  ii David, whose salary is $4000.
 	
  iii Riya, whose salary is $6000 with the lowest spent.
 	
  iv Sam, whose salary is $7000 with the highest spent.
	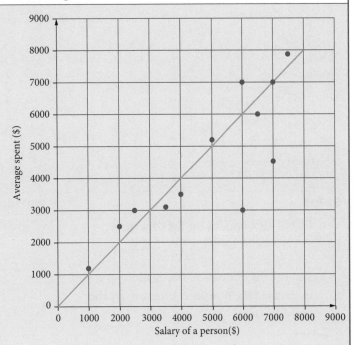

Read the vertical distance from the point to the line of best fit from the graph. The value is negative if it is below the line and zero if it is on the line. Include the response variable's units.	

b For a study done to establish the association between the height (cm) and weight (kg) of patients in a clinic, find the residual values for the following from the least squares line of best fit equation, correct to one decimal place:
 	
  $height = 80 + 1.1 \times weight$
 	
  i Maria who is 156 cm height and weighing 71 kg.
 	
  ii John who is 180 cm height and weighing 79 kg.	

1 Calculate the predicted height by substituting the weight into the line of best fit equation.	
2 Find the residual value using the formula. residual value = actual value – predicted value Include the response variable's units.	

SB

Using CAS 4:
Calculating residual values from a table
p. 169

SB

Using CAS 5:
Creating a residual plot
p. 172

p. 184

| MATCHED EXAMPLE 8 | Writing data transformation equations |

For each of the following non-linear scatterplots, write the squared, log and reciprocal transformed lines of best fit equations that linearise the data.

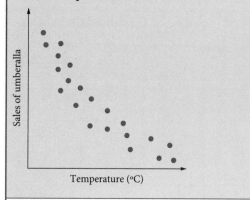

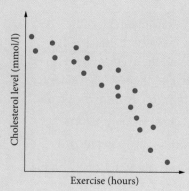

| Steps | Working |

a **1** Identify the options for the data shape.

2 Rewrite the line of best fit equation $y = a + bx$ in terms of the transformations, using the variable names.

b **1** Identify the options for the data shape.

2 Rewrite the line of best fit equation $y = a + bx$ in terms of the transformations, using the variable names.

SB

p. 185

The data for the association between the daily sales and the number of customers in a shop has been linearised by applying two separate transformations, giving the following least squares line of best fit equations.

$sale = 6 + 1.3 \times (number\ of\ customers)^2$

$\log(sale) = 3.1 + 0.01 \times (number\ of\ customers)$

a For each of these, use the equation to predict the sales when the number of customers is

  i 30   **ii** 55   **iii** 80

b What shape is the original data?

c The coefficient of determination was calculated to be 0.90 for the squared transformation and 0.85 for the log transformation. Which of the two equations gives the best fit to the data? Give a reason for your answer.

Steps	Working
a Substitute the value into the equation.	
Use CAS when dealing with logs.	
$\log_{10}(x) = a$ is the same as $x = 10^a$.	
$\log x$ means $\log_{10}(x)$.	
Round the answer if necessary.	
b 1 Which transformations have been used?	
2 Which data shape matches both of these transformations?	
c Which of the two coefficients of determination is closer to 1?	

SB

Using CAS 6A:
Transforming
non-linear data
with TI-Nspire
p. 186

SB

Using CAS 6B:
Transforming
non-linear data
with Casio
ClassPad
p. 188

CHAPTER

4 TIME SERIES

SB

Using CAS 1:
Constructing time
series plots
p. 209

SB

p. 211

MATCHED EXAMPLE 1 | Finding the median of a time series

This time series plot shows the number of audience members in a cinema each week over a twelve-week period. Find the median number of weekly audience members during this period.

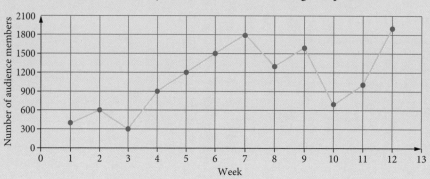

Steps	**Working**
1 Find the number of data values, *n* and note if the number is odd or even.	
2 Find the position of the median.	
3 Count up from the lowest to find the values. Read the values from the graph. If *n* is odd, find the middle value. If *n* is even, average the two middle values.	Use the above graph.

MATCHED EXAMPLE 2 | Identifying seasonality in time series

For each of the following, state whether a time series of the data is likely to show seasonality. Explain your reasoning.

a sales of washing machines

b families going for vacation

c sales of summer clothes

d sales of butter

Steps	Working
Is there likely to be a pattern throughout the year?	
If so, what time period would the pattern be based on?	

MATCHED EXAMPLE 3 | Moving mean smoothing with an odd number of points

The average mathematics test scores of Year 10 students over the last twelve years are shown in the following table.

Year	1	2	3	4	5	6	7	8	9	10	11	12
Marks scored	40	60	75	70	89	80	70	82	75	80	85	95

a Use a table with three columns to calculate the smoothed data using the method of three-point moving mean.

b What are the smoothed marks scored for the fifth and twelfth years?

c Graph the original data and the smoothed data on the same set of axes.

d What does the graph of the smoothed data indicate about the trend in the original data?

Steps	**Working**

a 1 Set up a table with three columns with Year in column 1, Marks scored in column 2 and the calculations for three-point moving means in column 3.

2 Find the mean for each group of three consecutive values for the number of students.

3 Write the mean in the row of the middle number of the three consecutive values.

Note: You cannot find a moving mean for the first and last value for the number of students.

Year	Marks scored	Three-point moving mean
1	40	
2	60	
3	75	
4	70	
5	89	
6	80	
7	70	
8	82	
9	75	
10	80	
11	85	
12	95	

b Read from the table.

c Graph the original data and the smoothed data on the same set of axes.

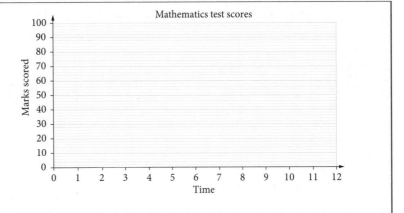

d Identify whether the smoothed graph is increasing or decreasing.

SB

p. 220

MATCHED EXAMPLE 4 | Moving mean smoothing with an even number of points

The following table shows the number of single-family housing permits in the year 2015 in Melbourne.

Month	Jan	Feb	Mar	Apr	May	Jun	Jul	Aug	Sep	Oct	Nov	Dec
Housing permits	7540	6005	6509	5890	4990	4500	5500	5200	4995	4378	3990	3000

a Use a table with four columns to calculate the smoothed data using the method of four-point moving means with centring.

b What is the smoothed housing permits for April and November?

c Graph the smoothed data.

d What does the graph of the smoothed data indicate about the trend in the original data?

e Find the smoothed data values for February and March using the **two**-point moving mean method with centring.

Steps | Working

a Set up a table with four columns and extra rows between time intervals for the centring of four-point moving means.

Month	Single-family housing permits	Four-point moving means	Four-point moving mean with centring
Jan	7540		
Feb	6005		
Mar	6509		
Apr	5890		
May	4990		
Jun	4500		
Jul	5500		
Aug	5200		

46 Nelson VICmaths General Mathematics 12 Mastery Workbook

9780170464055

Month	Single-family housing permits	Four-point moving means	Four-point moving mean with centring
Sep	4995		
Oct	4378		
Nov	3990		
Dec	3000		

b Read the values from the table.

c Graph the smoothed data by hand.

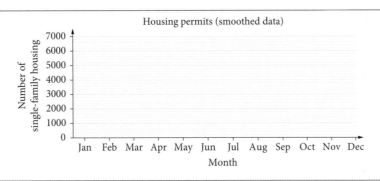

Housing permits (smoothed data)

d Identify from the graph.

e Set up a table with four columns and extra rows between time intervals for the centring of two-point moving means.

Read the values from the table.

Month	Single-family housing permits	Two-point moving means	Two-point moving mean with centring
Jan	7540		
Feb	6005		
Mar	6509		
Apr	5890		

The smoothed data value for February is 6515 and that for March is 6228.

MATCHED EXAMPLE 5 Graphical smoothing using moving medians

For this time series plot showing the marks scored by form 10 over the last 12 years, smooth the data using the

a three-point moving median method.

b five-point moving median method.

c What do the graphs of the smoothed data indicate about the trend in the original data?

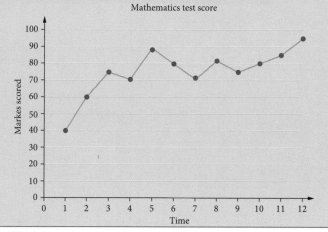

Steps	Working

a 1 Look at the first three points and find the middle point both left-to-right and bottom-to-top, giving you the coordinates of the first smoothed point.

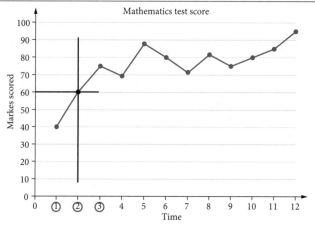

2 Look at the next three points and repeat.

Draw on the graph in step 1.

3 Repeat for the next three points. Sometimes the smoothed point is the same as the original point.

Draw on the graph in step 1.

4 Repeat the process until you reach the last median and then join the points.

Draw on the graph in step 1.

b 1 Look at the first five points and find the middle point both horizontally and vertically, giving you the coordinates of the first smoothed point.

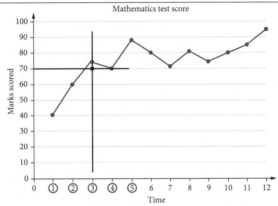

Mathematics test score

2 Look at the next five points and repeat.

Draw on the graph in step 1.

3 Repeat the process until you reach the last median and then join the points.

Draw on the graph in step 1.

c Is there an increasing or decreasing trend?

MATCHED EXAMPLE 6 | Interpreting seasonal indices

The following table shows the seasonal indices for the number of burgers sold on each day of the week by a burger shop:

Mon	Tue	Wed	Thu	Fri	Sat	Sun
1.01	1.3	0.41	0.3	1.33	1.05	1.6

Steps	**Working**
a How many days had below average sales?	
How many indices are less than 1?	
b Which day had the closest to average sales?	
Which of the indices is closest to 1?	
c Which day had sales furthest away from the average?	
Which of the indices is furthest from 1?	
d What do the seasonal indices add to?	
Add the indices.	
e What is the mean of the seasonal indices?	
Divide the sum of the indices by the number of indices.	
f Rewrite the table so that the seasonal indices are converted to percentages.	
Convert each decimal to a percentage.	

Mon	Tue	Wed	Thu	Fri	Sat	Sun

g What is the mean of the percentaged seasonal indices?	
Add the percentages and divide by 7.	
h What percentage below average were Thursday's sales?	
Compare the percentage to 100%.	

MATCHED EXAMPLE 7 | Calculating seasonal indices

The quarterly sales figures for the number of mobiles sold were recorded by a mobile company for 2020.

Quarter 1	Quarter 2	Quarter 3	Quarter 4
54	32	44	18

a Find the seasonal indices for each of the quarters, correct to two decimal places.

b Rewrite the table so that the seasonal indices are converted to percentages.

c Which of the quarters' sales were 50% below average?

Steps	Working

a 1 Find the average for the season.

2 For each of the actual figures, use the formula

$$\text{seasonal index} = \frac{\text{actual figure}}{\text{average for the season}}$$

rounding your answer to two decimal places.

Quarter 1	Quarter 2	Quarter 3	Quarter 4

b Convert each decimal to a percentage.

Quarter 1	Quarter 2	Quarter 3	Quarter 4

c Read from the percentage table.

SB

p. 238

MATCHED EXAMPLE 8 Dealing with seasonalised time series data

The quarterly sales figures for the number of mobiles sold were recorded by a mobile company from 2023 to 2025.

Year	Q1	Q2	Q3	Q4
2023	65	67	40	30
2024	45	80	50	65
2025	75	85	70	80

a Calculate the seasonal index for each quarter, correct to four decimal places.

b Use the seasonal indices to de-seasonalise the data, correct to two decimal places.

c Plot the original and the de-seasonalised data on the same set of axes.

Steps | **Working**

a 1 Calculate the yearly mean. Calculate the totals for each year, and find the mean for each year by dividing each total by 4.

Year	Q1	Q2	Q3	Q4	Yearly mean
2023	65	67	40	30	
2024	45	80	50	65	
2025	75	85	70	80	

2 Calculate the quarterly proportions. Divide each quarterly sales figure by the corresponding yearly mean to obtain quarterly proportions. Give answers correct to four decimal places.

Year	Q1	Q2	Q3	Q4
2023				
2024				
2025				

3 Calculate the seasonal indices by finding the mean of the quarterly proportions. Give answers correct to four decimal places.

4 Add the four seasonal index values to check the sum is 4.

5 Write a summary table for the seasonal indices.

Year	Q1	Q2	Q3	Q4
Seasonal index				

b 1 De-seasonalise the original time series data using the formula

de-seasonalised figure =

$$\frac{\text{actual figure}}{\text{seasonal index}}$$

Write your answers in the table below.

Year	Q1	Q2	Q3	Q4
2023				
2024				
2025				

2 Write a summary table for the de-seasonalised values, correct to two decimal places.

Year	Q1	Q2	Q3	Q4
2023				
2024				
2025				

c Plot the original and the de-seasonalised time series data on the same set of axes.

Use $t = 1$ to represent the first quarter of 2023, $t = 2$ to represent the second quarter of 2023 etc.

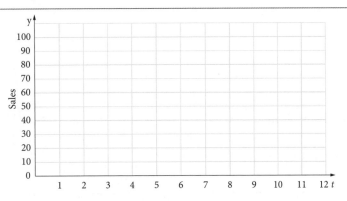

MATCHED EXAMPLE 9 | De-seasonalising and re-seasonalising time series data

The following table shows the quarterly seasonal indices for the sales of a motorcycle company.

Quarter	1	2	3	4
Seasonal index	1.1	0.7	1.3	

a What is the missing quarter 4 seasonal index?

b To correct for seasonality, by what percentage should the sales for quarter 4 be increased?

c The company predicts that its de-seasonalised quarterly sales will be $5 000 000 for each quarter. Based on this, what would you predict the actual sales for quarter 4 to be?

Steps	Working
a Use the fact that the seasonal indices need to add to 4 for quarterly data.	
b Use the de-seasonalising formula $$\text{de-seasonalised figure} = \frac{\text{actual figure}}{\text{seasonal index}}$$	
c Use the re-seasonalising formula actual figure = de-seasonalised figure $\times$ seasonal index	

MATCHED EXAMPLE 10 | Working with trend lines for de-seasonalised data

The following table lists the de-seasonalised number of sales of a particular ring in a jewellery store for each quarter in 2022–2023 and the seasonal indices.

Quarter	1	2	3	4
De-seasonalised number of sales in 2022	5	11	9	15
De-seasonalised number of sales in 2023	22	29	24	28
Seasonal index	1.4	0.9	0.6	1.1

a Find the equation of the least squares trend line for the de-seasonalised time series data for 2022–2023. Round the slope and intercept to three significant figures.

b Plot the time series and draw the trend line for the de-seasonalised data on the same axes, and comment on the trend by interpreting the slope of the trend line equation.

c Use the trend line equation to forecast the de-seasonalised number of sales for quarter 2 2024.

d Use the trend line equation to forecast the actual number of sales for quarter 2 2024.

Steps	Working

a 1 Rewrite the de-seasonalised number of sales in a table that represents the quarters from 1 to 8.

2 Use CAS to find a and b for the least squares line of best fit equation, rounding to the required significant figures.

Quarter number	1	2	3	4	5	6	7	8
De-seasonalised number of sales 2022–2023								

TI-Nspire

1 Open a **Lists & Spreadsheet** page.

2 Enter the appropriate headings and the values in columns **A** and **B**.

3 Press **menu > Statistics > Stat Calculations > Linear Regression (a+bx)**.

4 Select the headings then select **OK**.

1 Tap **Menu > Statistics**.

2 Enter the values into **list1** and **list2**.

3 Tap **Calc > Regression > Linear Reg**.

4 On the next screen, keep the default settings of **XList: list1** and **YList: list2** and tap **OK**.

5 Select **y=a+bx** from the dropdown menu.

b Use CAS to plot the time series and draw the line of best fit.

Refer to the slope from the trend line equation in your comment.

TI-Nspire

1 Add a **Data & Statistics** page.

2 Select the headings from the **Lists & Spreadsheet** columns.

3 Press **menu > Plot Properties > Connect Data Points**.

4 Press **menu > Analyze > Regression > Show Linear (a+bx)**.

5 The time series and trend line will both be displayed.

1 Tap OK from the previous screen to display the scatterplot and least squares regression line in the lower window.

2 Tap the **Set StatGraphs** tool.

3 Change the **Type:** from **Scatter** to **xyLine** and tap Set.

4 The time series and trend line will both be displayed.

5 Tap the **Equation** tool to display the equation.

c Find the number of the quarter and use the trend line equation to forecast the de-seasonalised number of sales.

Round the answer to whole rings sold.

d Re-seasonalise by using

actual figure = de-seasonalised figure × seasonal index

Use the **unrounded** de-seasonalised figure, then round the answer to whole rings sold.

CHAPTER

5 INTEREST AND DEPRECIATION

SB

p. 268

MATCHED EXAMPLE 1	Generating a sequence using a recurrence relation

A sequence has the recurrence relation $u_0 = 1$, $u_{n+1} = 2u_n + 2$.

a Describe in words the calculations required to generate the sequence.

b Find the first four terms of the sequence generated by this recurrence relation, showing all the steps.

c Find u_5.

Steps	Working
a State the starting value and the rule for the recursive relation.	
b **1** Write the first value u_0.	
2 Substitute u_0 into the rule to find u_1.	
3 Repeat for the next two values, u_2 and u_3.	
4 List the first four terms of the sequence.	
c Use the rule to find u_4 first and then use it to find u_5.	

SB

Using CAS 1:
Generating
a sequence
using recursive
computation
p. 269

MATCHED EXAMPLE 2	Writing recurrence relations

Write the recurrence relation for each of the following.

a Start with 2. Divide each value by 2 and then subtract 0.25 to find the next value.

b 6, 30, 150, 750, …

c −10, −15, −20, −25, …

Steps	Working
a What is the starting value? What is the rule connecting each value to the value before it?	
b 1 What is the starting value? **2** What is the rule connecting each value to the value before it? **3** Write the recurrence relation.	
c 1 What is the starting value? **2** What is the rule connecting each value to the value before it? **3** Write the recurrence relation.	

5

MATCHED EXAMPLE 3 | Identifying graphs of recurrence relations

State whether each of the following is true or false and give a reason for your answer in each case.

a The graph of the recurrence relation $V_0 = 3000$, $V_{n+1} = V_n + 300$ consists of points in an increasing straight line.

b The graph of the recurrence relation $V_0 = 200$, $V_{n+1} = V_n - 30$ consists of points in an increasing straight line.

c The graph of the recurrence relation $V_0 = 10\,000$, $V_{n+1} = 5Vn$ will look like

d The following graph of a recurrence relation shows geometric growth.

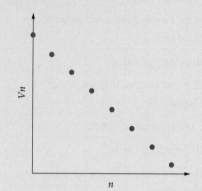

Steps	Working
a Is addition or subtraction involved? Is V_n multiplied by a number greater than 1 or by a number between 0 and 1?	
b Is addition or subtraction involved? Is V_n multiplied by a number greater than 1 or by a number between 0 and 1?	
c Is addition or subtraction involved? Is V_n multiplied by a number greater than 1 or by a number between 0 and 1?	
d Is the graph a straight line or a curve? Is the graph increasing or decreasing?	

MATCHED EXAMPLE 4 | Using simple interest recurrence relations

Oliver invests $4000 in a bank account earning 2% per annum simple interest.

p. 277

Steps	Working

a What is the fixed amount of interest paid for each year?

Use $d = \dfrac{r}{100} \times V_0$ to find the fixed

amount of interest for each year.

b Copy and complete the following table to find

  i Oliver's bank account balance after four years

  ii the first year when his balance is greater than $4300

  iii the total amount of interest earned after six years.

n	Account balance after n years ($)
0	4000
1	4000 + =
2	+ =
3	+ =
4	+ =
5	+ =
6	+ =

Complete the table by using CAS recursive computation to find the bank account balance after six years.

Enter your answers in the above table.

  i Read the answer from the table.

  ii Read the answer from the table.

  iii Total amount of interest earned
     on an investment after n years
     $= V_n - V_0$

c Write a recurrence relation for the account balance.

Identify the starting value. Each
value is calculated by adding d to the
previous value.

d Describe the sort of growth or decay modelled by the recurrence relation.

Is there addition or subtraction
involved?

Is there multiplication involved by a
number greater than 1 or between 0
and 1?

e Sketch the graph of the recurrence relation up to $n = 6$.

The horizontal axis is n (years) and
the vertical axis is V_n ($).

Plot the values from the table.

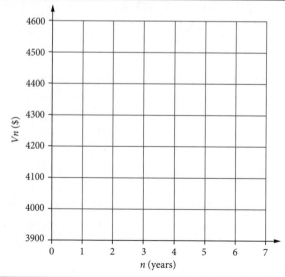

MATCHED EXAMPLE 5 | Using recursive computation for simple interest

Find the balance after six years for each of the following using CAS recursive computation.

a Jack invests $5000 in a bank account earning 3% per annum simple interest.

b Anna takes out a loan of $5000 from a bank at 3% per annum simple interest.

Steps	Working
a 1 Use $d = \dfrac{r}{100} \times V_0$ to find the fixed amount of interest for each year. Add d for an investment and subtract d for a loan.	
2 Using CAS recursive computation to find the balance after six years.	

TI-Nspire   **ClassPad**

3 Write the answer.

b 1 Use $d = \dfrac{r}{100} \times V_0$ to find the fixed amount of interest for each year.

Add d for an investment and subtract d for a loan.

2 Using CAS recursive computation to find the
balance after six years.

3 Write the answer.

MATCHED EXAMPLE 6 | Using the simple interest general rule

Ellyse invests $10 000 in an account earning 5% p.a. simple interest.

Steps	Working
a Find the fixed amount of interest paid each year.	
Find the value of d.	
b Write a rule that will calculate the balance of the account after n years.	
Substitute the values of d and V_0 into the simple interest general rule. Decide if it is an investment or loan.	
c Use the rule to find the balance of the account after eight years.	
Substitute the value of n into your rule.	
d Use the rule to find how many years it would take for the investment to double.	
1 Substitute the known values into $V_n = V_0 + nd.$	
2 Solve for n, using CAS if necessary. Round *up* to the nearest year if necessary.	

TI-Nspire   ClassPad

3 Write the answer.

SB

p. 285

MATCHED EXAMPLE 7 | Using flat rate depreciation recurrence relations

A mobile phone is purchased for $1550. Its value depreciates at a flat rate of 20% each year.

Steps	Working

a What is the fixed amount of depreciation each year?

Use $d = \dfrac{r}{100} \times V_0$ to find the fixed
amount of depreciation each year.

b Copy and complete the following table to find

  i the value of the mobile phone after four years

  ii when the value of the mobile phone first falls below $700

  iii when the mobile phone depreciates to zero.

n	Value after n years ($)
0	1550
1	1550 – =
2	– =
3	– =
4	– =
5	– =
6	– =

  1 Complete the table by using CAS recursive computation to find the value of the asset after five years.

Enter your answers in the above table.

  2 i Read the answer from the table.

    ii Read the answer from the table.

    iii Read the answer from the table.

c Write a recurrence relation for the value of the mobile phone.

Identify the initial value of the asset.
Each value is calculated by subtracting
d from the previous value.

d Describe the sort of growth or decay modelled by the recurrence relation.

Is there addition or subtraction
involved?

Is there multiplication involved by a
number greater than 1 or between
0 and 1?

e Sketch the graph of the recurrence relation up to $n = 5$.

The horizontal axis is n (years) and the vertical axis is V_n ($).

Plot the values from the table.

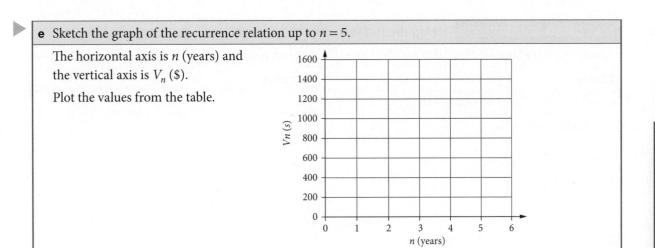

MATCHED EXAMPLE 8 | Using the flat rate depreciation general rule

A business buys a boat for $900 000 and the value of the boat depreciates at a flat rate of 10% per year.

Steps	Working
a Find the fixed amount of depreciation each year.	
Find the value of d.	
b Write a rule that will calculate the value of the boat after n years.	
Substitute the values of d and V_0 into the flat rate depreciation general rule.	
c Use the rule to find the value of the boat after six years.	
Substitute the value of n into your rule.	
d Use the rule to find how many years it would take for the boat to depreciate to zero.	

1 Substitute the known values into
     $V_n = V_0 - nd$.
     Let $V_n = 0$.	

2 Solve for n, using CAS if necessary. Round *up* to the nearest year if necessary.	
3 Write the answer.	

MATCHED EXAMPLE 9 | Finding the rate for flat rate depreciation

Anna paid $3000 for her new bridal gown. After three years of flat rate depreciation, the resale value of the gown was $1650.

a What is the fixed amount of depreciation each year?

b What was the annual flat rate of depreciation she used, as a percentage of the purchase price?

Steps	Working
a 1 Identify what we know and what we need to find from the general rule for flat rate depreciation after n years.	
2 Substitute into the rule and solve, using CAS if necessary.	
3 Write the answer.	
b 1 Identify what we know and what we need to find.	
2 Substitute the values into $$r = \frac{d}{V_0} \times 100\%$$ and evaluate.	
3 Write the answer.	

MATCHED EXAMPLE 10 Using unit cost depreciation recurrence relations

A video gamer buys the Super Mario 64 game in an auction for $200 000, which depreciates by $20 000 every time it's played.

Steps	Working

a Explain why this is unit cost depreciation and not flat rate depreciation.

 1 Refer to 'use' in your answer.

b Copy and complete the following table to find

 i the value of the game after four plays.

 ii how many plays will it take for the value of the game to first fall below $120 000.

n	Value after n units of use ($)
0	200 000
1	200 000 – =
2	– =
3	– =
4	– =
5	– =

 1 Complete the table by using CAS recursive computation to find the value after five plays.

Enter your answers in the above table.

 2 i Read the answer from the table.

 ii Read the answer from the table.

c Write a recurrence relation for the value of the game.

Identify the initial value of the asset and the cost per unit of use.

d Describe the sort of growth or decay modelled by the recurrence relation.

Is there addition or subtraction involved?

Is there multiplication involved by a number greater than 1 or between 0 and 1?

e Sketch the graph of the recurrence relation up to $n = 5$.

The horizontal axis is n (units of use) and the vertical axis is V_n ($).

Plot the values from the table.

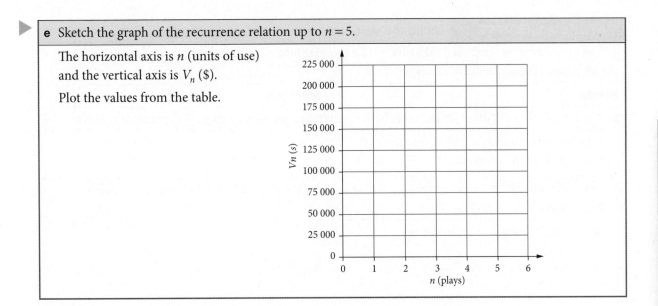

MATCHED EXAMPLE 11 Using the unit cost depreciation general rule

A concrete mixer was purchased for $70 000. The concrete mixer's value depreciates at a rate of 80 cents per kilogram of concrete that it mixes.

Steps	Working
a Write a rule that will calculate the value of the concrete mixer after mixing n kilograms of concrete.	
Substitute the values of d and V_0 into the unit cost depreciation general rule $V_n = V_0 - nd$.	
b Use the rule to find the value of the concrete mixer after it has mixed a total of 60 000 kilograms of concrete.	
Substitute the value of n into your rule.	
c How many kilograms has the concrete mixer mixed when it depreciates to $30 000? Give your answer to the nearest kilogram.	
Use the given value of V_n and solve the rule for n, using CAS if necessary. Round *up* to the nearest whole number.	
d How many kilograms has the concrete mixer mixed when it depreciates to zero? Give your answer to the nearest kilogram.	
Use the given value of V_n and solve the rule for n, using CAS if necessary. Round *up* to the nearest whole number.	

9780170464055

MATCHED EXAMPLE 12 | Finding the cost per unit of use

The purchase price of a water bottle labelling machine is $12 000. After four years, the labelling machine has a value of $2024. On average, 50 600 water bottles were labelled every year during those four years. The value of the labelling machine was depreciated using a unit cost method of depreciation. Find

a the depreciation in the value of the labelling machine, per bottle labelled to the nearest cent.

b the value of the labelling machine after n units are produced, V_n.

Steps	**Working**
a 1 Find n, the number of units of use during the time period. Note n is *not* the number of years.	
2 Identify what we know and what we need to find from the general rule for unit cost depreciation after n units of use.	
3 Substitute into the rule and solve, using CAS if necessary.	
4 Write the answer to the nearest cent.	
b Substitute the values of d and V_0 into the unit cost depreciation general rule $V_n = V_0 - nd$ and simplify.	

MATCHED EXAMPLE 13 Comparing compound and simple interest

David is investing $4000 for five years and wants to compare an investment at 4% p.a. compounding yearly to 4% p.a. simple interest.

a Copy and complete the following table for $n = 3$, $n = 4$ and $n = 5$. Round your answers to the nearest cent.

	Compound		Simple	
n	Interest ($)	Value of investment ($)	Interest ($)	Value of investment ($)
0	–	4000	–	4000
1	$\frac{4}{100} \times 4000 = 160$	$4000 + 160 = 4160$	$\frac{4}{100} \times 4000 = 160$	$4000 + 160 = 4160$
2	$\frac{4}{100} \times 4160 = 166.40$	$4160 + 166.40 = 4326.40$	$\frac{4}{100} \times 4000 = 160$	$4160 + 160 = 4320$
3				
4				
5				

b What is the value of the compound interest investment after five years?

c After five years, how much more is the value of the compound interest investment compared with the simple interest investmen?

Steps	Working
a Enter your answers in the above table.	
b Read from the table.	
c Compare the last entries in the two Value of investment ($) columns in the table.	

MATCHED EXAMPLE 14 | Working with compounding periods

SB

p. 299

For each of the following investments find

    i the number of compounding periods per year

    ii the number of compounding periods over 10 years

    iii The percentage interest rate per compounding period

    iv The amount of interest earned in the first compounding period to the nearest cent.

a Amelia invests $10 000 at 10% compound interest per annum compounding quarterly.

b William invests $80 000 at 5% compound interest per annum compounding monthly.

Steps	Working
a **i** How many compounding periods are there per year?	
ii Multiply the number of compounding periods per year by the number of years.	
iii Divide the percentage interest rate per year by the number of compounding periods per year.	
iv Convert the compounding period interest rate to a decimal and multiply by the principal. Round to the nearest cent.	
b **i** How many compounding periods are there per year?	
ii Multiply the number of compounding periods per year by the number of years.	
iii Divide the percentage interest rate per year by the number of compounding periods per year.	
iv Convert the compounding period interest rate to a decimal and multiply by the principal. Round to the nearest cent.	

| MATCHED EXAMPLE 15 | Finding compound interest recurrence relations |

Write a recurrence relation for the account balance for each of the following savings accounts earning compound interest.

a Zoe deposited $5000 into an account at the rate of 4% per annum, compounding annually.

b Sophia deposited $8000 into an account at the rate of 3.6% per annum, compounding monthly.

c Chloe deposited $10 000 into an account at the rate of 6.5% per annum, compounding weekly.

d Describe the sort of growth or decay modelled by the recurrence relations.

e Sketch the shape of the graph of the recurrence relation.

Steps	Working
a 1 Find the number of compounding periods per year.	
2 Identify V_n, V_0 and r.	
3 Substitute the values into $V_0 =$ principal, $V_{n+1} = \left(1 + \dfrac{r}{100}\right) V_n$ and simplify.	
b 1 Find the number of compounding periods per year.	
2 Identify V_n, V_0 and r.	
3 Substitute the values into $V_0 =$ principal, $V_{n+1} = \left(1 + \dfrac{r}{100}\right) V_n$ and simplify.	
c 1 Find the number of compounding periods per year.	
2 Identify V_n, V_0 and r.	
3 Substitute the values into $V_0 =$ principal, $V_{n+1} = \left(1 + \dfrac{r}{100}\right) V_n$ and simplify.	

d Is there addition or subtraction involved?

Is there multiplication involved by a number greater than 1 or between 0 and 1?

e Show the points forming the shape of the curve.

Vn

n

MATCHED EXAMPLE 16	Working with compound interest recurrence relations

Grace has invested her money in an account earning compound interest, compounding annually, according to the recurrence relation

$$V_0 = 10\,000, \quad V_{n+1} = 1.05 V_n$$

where V_n is the account balance after n compounding periods.

Steps	Working
a How much money did Grace initially invest?	
Identify V_0.	
b Use recursion to write down calculations that show that the amount of money in Grace's account after three years will be \$11 576.25.	
Step out the recurrence relation calculations to find V_3.	
c What is the annual percentage compound interest rate for this account?	
1 Use the recurrence relation to find an equation for r, the percentage interest rate per compounding period.	
2 Solve for r, using CAS if necessary.	
3 Multiply r by the compounding period to find the annual percentage compound interest rate.	
d After how many years will the balance of Grace's account first exceed \$13 000?	
Use CAS recursive computation to continue the recurrence relation calculations until they are greater than the given amount.	

9780170464055

e What is the total interest, to the nearest cent, earned by Grace's investment after six years?

Total amount of interest earned after n
compounding periods $= V_n - V_0$.

Use CAS recursive computation to find V_n.

Round your answer to the nearest cent.

SB

Using CAS 3:
Creating interest
graphs
p. 303

SB

p. 306

MATCHED EXAMPLE 17	Using the compound interest rule

Kiara invests $25 000 in a bank where she earns interest of 3.6% p.a. compounded monthly.

Steps	Working
a Find r, the percentage interest rate per compounding period.	
Divide the yearly interest rate by the number of compounding periods per year.	
b Write a rule that will calculate the value of the investment after n months.	
Substitute the values of V_0 and r into the compound interest general rule $V_n = \left(1 + \dfrac{r}{100}\right)^n \times V_0$ and simplify.	
c Use the rule to find the value of the investment after five years to the nearest cent.	
1 Find the value of n, the number of compounding periods.	
2 Substitute the value of n, the number of compounding periods, into your rule.	
3 Write the answer, rounding to the nearest cent.	
d Use a rule to find the value of Kiara's investment after five years to the nearest cent if the interest was compounded quarterly instead of monthly.	
1 Find r by dividing the yearly interest rate by the number of compounding periods per year.	
2 Find the value of n, the number of compounding periods.	
3 Substitute n, r and V_0 into the rule.	
4 Write the answer, rounding to the nearest cent.	
e Which compounding period gives the larger balance after five years?	
Compare the two values.	

9780170464055

MATCHED EXAMPLE 18	Working with the compound interest rule

Find each of the following using the compound interest rule.

SB

p. 307

Steps	**Working**

a Amy invested \$12 000 in an account earning interest compounding annually, and after six years, the balance is \$17 215.85. What was the annual interest rate of the account to one decimal place?

1 Identify what we know and what we need to find from the compound interest rule.

2 Substitute into the rule and solve using CAS. Ignore the negative solution.

TI-Nspire   **ClassPad**

3 Write the rounded answer.

b Olivia invested \$5500 in an account earning 10% interest per annum compounding quarterly. How many quarters will it take to grow to \$10 000?

1 Identify what we know and what we need to find from the compound interest rule.

2 Substitute into the rule and solve using CAS.

TI-Nspire   **ClassPad**

3 Write the answer rounded *up* to the nearest quarter.

c Hugo has a compound interest investment that earns 6% interest compounding monthly. If the balance of Hugo's investment was \$20 232.75 after five years, what amount did Hugo initially invest to the nearest dollar?

1 Identify what we know and what we need to find from the compound interest rule.

2 Substitute into the rule and solve using CAS.

TI-Nspire

ClassPad

3 Write the rounded answer.

MATCHED EXAMPLE 19 | Finding effective interest rates

Valerie is looking to invest her money. She has done some research on interest rates and found the best offers from four different banks.

Bank 1: 5.5% p.a. compounding daily

Bank 2: 5.45% p.a. compounding monthly

Bank 3: 5.8% p.a. compounding six-monthly

Bank 4: 5.85% p.a. compounding annually

a Find the effective interest rate for each bank, rounding to two decimal places.

b Which bank should Valerie choose if she wants to earn the most interest?

c Which bank would earn Valerie the least interest?

d Why are the nominal and effective interest rates for Bank 4 the same?

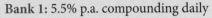

p. 312

Steps	Working
a For each option, substitute the known variables into the effective interest rate formula, rounding to two decimal places. $$r_{\text{effective}} = \left[\left(1 + \frac{r}{100n}\right)^n - 1\right] \times 100\%$$	
b Compare the four results and choose the largest.	
c Compare the four results and choose the smallest.	
d Compare the nominal and effective interest rates.	

Using CAS 4:
Finding effective
interest rates
p. 313

MATCHED EXAMPLE 20 | Using reducing balance depreciation recurrence relations

A business purchased a truck for $65 000. It is depreciated using reducing balance depreciation at a rate of 20% per annum. Give all answers to the nearest dollar.

Steps	Working

a Copy and complete the table to find

  i the value of the truck after six years

  ii the amount of depreciation in the fourth year

  iii when the truck first depreciates to under $25 000.

n	Depreciation after n years ($)	Value after n years ($)
0		65 000
1	$\frac{20}{100} \times 65\,000 = 13\,000$	$65\,000 - 13\,000 = 52\,000$
2	$\frac{20}{100} \times 52\,000 = 10\,400$	$52\,000 - 10\,400 = 41\,600$
3		
4		
5		
6		

a Calculate the percentage of successive values and subtract from the previous value.

Use CAS's recursive computation where possible.

Give all values to the nearest dollar, but don't round until after all the calculations have been done.

(Note: Answers can vary slightly depending on when values are rounded.)

Enter your answers in the above table.

  i Read the answer from the table.

  ii Read the answer from the table.

  iii Read the answer from the table.

b Write down a recurrence relation that gives the value of the truck after n years.

> **1** Identify V_n, V_0 and r.
>
> **2** Substitute the values into
>
> V_0 = initial value of the asset,
>
> $$V_{n+1} = \left(1 - \frac{r}{100}\right) V_n$$
>
> and simplify.

c What percentage of the previous value is each new value?

> Look at the decimal in front of V_n and convert it to a percentage.

d Describe the sort of growth or decay modelled by the recurrence relation.

> Is there addition or subtraction involved?
>
> Is there multiplication involved by a number greater than 1 or between 0 and 1?

e Use the graph to find the truck's approximate value after eight years.

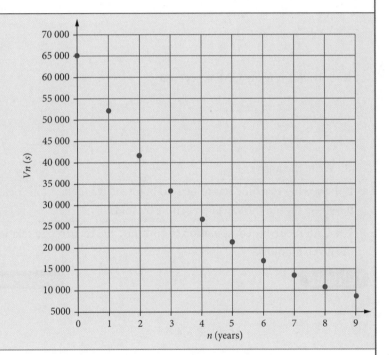

> Read from the graph.

MATCHED EXAMPLE 21	Working with reducing balance depreciation recurrence relations

The value of a sewing machine is depreciated using the reducing balance method. A recurrence relation modelled to determine the value of the sewing machine, V_n, in dollars, after n years is given by

$$V_0 = 600, \qquad V_{n+1} = 0.9V_n$$

Steps	Working

a Use recursion to show that the value of the sewing machine after two years, V_2, is $486.

Step out the recurrence relation working to find V_2.

b What is the annual percentage rate of depreciation of the sewing machine?

1 Use the recurrence relation to find an equation for r.

2 Solve for r, using CAS if necessary.

3 Write the answer.

c If the sewing machine has to be sold before its value first falls below $300, how many years after its purchase should the sewing machine be sold?

Use CAS recursive computation to continue the recurrence relation calculations until they are less than the given amount.

TI-Nspire   **ClassPad**

A refrigerator was bought for $2600 and is being depreciated on a reducing balance basis rate of 10% per annum.

SB

p. 319

Steps	Working
a Write a rule that will calculate the value of the refrigerator after n years.	
Substitute the values of V_0 and r into the reducing balance depreciation general rule and simplify.	
b Use the rule to find the value of the refrigerator after 11 years to the nearest dollar.	
Substitute the value of n into your rule and solve.	
c Use the rule to find how many years it would take for the refrigerator to depreciate to under $500.	
1 Identify what we know and what we need to find from the reducing balance depreciation rule.	
2 Substitute into the rule and solve using CAS.	
TI-Nspire	ClassPad
3 Write the answer, rounding *up* to the nearest year.	
d How much is the refrigerator depreciated by in the fourth year to the nearest dollar?	
Amount of depreciation in the nth year = $V_{n-1} - V_n$.	

MATCHED EXAMPLE 23 | Working with the reducing balance depreciation rule

Find each of the following using the reducing balance depreciation rule.

Steps	Working

a Zola bought a boat for $60 000. After four years of being depreciated on a reducing balance basis, the boat is now valued at $39 366. Show that the annual rate of depreciation in the value of the boat is 10%.

1 Identify what we know and what we need to find from the reducing balance depreciation rule.

2 Substitute into the rule and show the steps to solve for r.

If this wasn't a 'show' question, we could use CAS.

b A construction company uses the reducing balance method to depreciate a bulldozer it owns. The annual rate of depreciation in the value of the bulldozer is 15%. If after eight years the value of the bulldozer is $107 633.76, what was the original price of the bulldozer to the nearest dollar?

1 Identify what we know and what we need to find from the reducing balance depreciation rule.

2 Substitute into the rule and solve using CAS.

TI-Nspire   ClassPad

3 Write the rounded answer.

CHAPTER 6

LOANS, INVESTMENTS AND FINANCE SOLVERS

MATCHED EXAMPLE 1	Using finance solvers for compound interest investments

Alexa invested \$65 000 in a term deposit earning 6.2% p.a. compounding yearly. Give answers to the nearest cent.

p. 338

Steps	**Working**
a What is the value of Alexa's investment after 10 years?	
	TI-Nspire + **ClassPad**
1 Find FV after $10 \times 1 = 10$ years.	
2 Total number of compounding periods	N
Annual interest rate	I%
Present value for an investment is negative.	PV
Payment amount	Pmt or PMT
Future value has the opposite sign to present value.	FV
Same as CpY or C/Y	PpY or P/Y
Number of compounding periods per year	CpY or C/Y
3 Write the answer, rounding to nearest cent.	
b How long will it take for her investment to grow to \$95 000?	
1 Enter the amount as FV and find N.	
2 Total number of compounding periods	N
Annual interest rate	I%
Present value for an investment is negative.	PV
Payment amount	Pmt or PMT
Future value has the opposite sign to present value.	FV
Same as CpY or C/Y	PpY or P/Y
Number of compounding periods per year	CpY or C/Y
3 Write the answer, rounding up to the next whole number.	
c i How long will it take for her investment to grow to \$95 000 if the interest is compounded monthly?	
** ii** How much longer will it take for yearly compounding to reach \$95 000 compared to monthly compounding?	

1 Change CpY or C/Y to compounding monthly and find N.

2

Total number of compounding periods	N
Annual interest rate	I%
Present value for an investment is negative.	PV
Payment amount	Pmt or PMT
Future value has the opposite sign to present value.	FV
Same as CpY or C/Y	PpY or P/Y
Number of compounding periods per year	CpY or C/Y

3 **i** Write the answer, rounding up to the next whole number.

4 **ii** Use the fact that there are 12 months in a year to compare.

9780170464055

MATCHED EXAMPLE 2	Using finance solvers for reducing balance depreciation

Oliver has been depreciating his office furniture on a reducing balance basis rate per annum.

Steps	**Working**

a He bought a couch for $800 and after three years its value is $540. What is the rate of depreciation, rounded to one decimal point?

TI-Nspire + **ClassPad**

1 Find I from negative PV to positive FV.

2 Total number of compounding periods — N

Annual interest rate for depreciation is negative. — I%

Present value for depreciation is negative. — PV

Payment amount — Pmt or PMT

Future value has the opposite sign to present value. — FV

Same as CpY or C/Y — PpY or P/Y

Number of compounding periods per year — CpY or C/Y

3 Write the answer, rounding to one decimal point.

b He has been depreciating his office desk at a rate of 6%. After three years, its value was $500. What was the office desk originally bought for to the nearest dollar?

1 Enter the amount as FV and find PV.

2 Total number of compounding periods — N

Annual interest rate for depreciation is negative. — I%

Present value for depreciation is negative. — PV

Payment amount — Pmt or PMT

Future value has the opposite sign to present value. — FV

Same as CpY or C/Y — PpY or P/Y

Number of compounding periods per year — CpY or C/Y

3 Write the answer, rounding to the nearest dollar.

MATCHED EXAMPLE 3	Using finance solvers for two-step compound interest investment problems

Solve each of the following by using a finance solver twice.

Steps	**Working**
a The balance of Steve's investment, which compounds quarterly, was \$10 500 after two quarters and \$11 105 after a year. What was the amount of money that he initially invested to the nearest dollar?	

TI-Nspire + ClassPad

1 Find the interest rate from negative PV to positive FV.	
2 Total number of compounding periods = 4 − 2 = 2 quarters	N
Annual interest rate	I%
Present value for an investment is negative.	PV
Payment amount	Pmt or PMT
Future value has the opposite sign to present value.	FV
Same as CpY or C/Y	PpY or P/Y
Number of compounding periods per year	CpY or C/Y
3 Use the unrounded answer to find the initial amount.	
4 Total number of compounding periods	N
Annual interest rate from previous solver (unrounded)	I%
Present value for an investment is negative.	PV
Payment amount	Pmt or PMT
Future value has the opposite sign to present value.	FV
Same as CpY or C/Y	PpY or P/Y
Number of compounding periods per year	CpY or C/Y
5 Write the answer, rounding to the nearest dollar.	

b The graph shows the value, in dollars, of a compound interest investment after n compounding periods, V_n, for a period of four compounding periods. What is the value of V_4, rounded to the nearest cent?

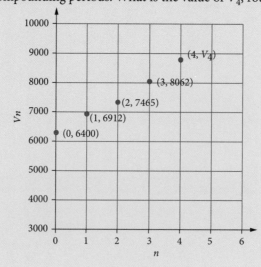

1 Find the interest rate using the value for $n = 1$ as FV.

2 Total number of compounding periods   N

   Annual interest rate   I%

   Present value for an investment is negative.   PV

   Payment amount   Pmt or PMT

   Future value has the opposite sign to present value.   FV

   Same as CpY or C/Y   PpY or P/Y

   Number of compounding periods per year   CpY or C/Y

3 Use the unrounded answer to find the value for $n = 4$.

4 Total number of compounding periods   N

   Annual interest rate from previous solver (unrounded)   I%

   Present value for an investment is negative.   PV

   Payment amount   Pmt or PMT

   Future value has the opposite sign to present value.   FV

   Same as CpY or C/Y   PpY or P/Y

   Number of compounding periods per year   CpY or C/Y

5 Write the answer, rounding to the nearest cent.

p. 344

MATCHED EXAMPLE 4 | Finding reducing balance loan recurrence relations

Michelle has taken out a loan of $25 000 with an interest rate of 10.5% per annum compounding quarterly, and she makes regular quarterly payments of $1580.

a Write a recurrence relation for the balance.

b Describe the sort of growth or decay modelled by the recurrence relation.

c Sketch the shape of the graph of the recurrence relation.

Steps	Working
a 1 Find the number of compounding periods per year.	
2 Identify V_n, V_0, r and d.	
3 Substitute the values into these formulas and simplify. $V_0 = \text{principal}$ $V_{n+1} = \left(1 + \dfrac{r}{100}\right)V_n - d$	
b Is there addition or subtraction involved? Is there multiplication involved by a number greater than 1 or between 0 and 1?	
c Show the points forming the shape of the curve.	V_n n

MATCHED EXAMPLE 5 | Working with reducing balance loan recurrence relations

Bruce has taken out a reducing balance loan, compounding monthly and with regular monthly payments, according to the recurrence relation

$$V_0 = 8000, \quad V_{n+1} = 1.0085V_n - 250,$$

where V_n is the value of the loan after n compounding periods.

Steps	Working
a How much money did Bruce borrow?	
Identify V_0.	
b How much are the regular monthly payments?	
Identify d.	
c Use recursion to write down calculations that show that the amount owing on his loan after three months will be $7449 when rounded to the nearest dollar.	
Step out the recurrence relation calculations to find V_3, giving answers to the nearest dollar.	
d What is the annual percentage compound interest rate for this loan?	
1 Use the recurrence relation to find an equation for r, the percentage interest rate per compounding period.	
2 Solve for r, using CAS if necessary.	
3 Multiply r by the compounding period to find the annual percentage compound interest rate.	

6

e After how many months will the balance of Bruce's loan fall below $6600?

Use CAS recursive computation to continue the recurrence relation calculations until they are less than the given amount.

TI-Nspire

ClassPad

9780170464055

MATCHED EXAMPLE 6 | Using reducing balance loan amortisation tables

Cersei has taken out a construction loan of $310 000 with interest compounding quarterly and is paying quarterly instalments of $9220. The amortisation table below shows the first few calculations for her loan.

Payment number	Payment	Interest	Principal reduction	Balance
0	0.00	0.00	0.00	310 000.00
1	9220.00	$\frac{r}{100} \times 310\,000.00$ $= 7068.00$	$9220.00 - 7068.00$ $= 2152.00$	$310\,000.00 - 2152.00$ $= 307\,848.00$
2	9220.00	$\frac{r}{100} \times 307\,848.00 = 7018.93$		
3				
4				

Steps | Working

a Use the table to show two calculations that will give r, the percentage interest rate per compounding period.

Use this formula:

$$r = \frac{\text{interest}}{\text{previous balance}} \times 100$$

b What is the nominal interest rate for the loan?

Use the compounding period to find the annual interest rate.

c Complete the amortisation table, showing the interest paid, principal reduction and balance for the first four quarters of the loan, giving all values to the nearest cent.

Complete the table using:

$\text{interest} = \frac{r}{100} \times \text{previous balance}$

$\text{principal reduction} = \text{payment} - \text{interest}$

$\text{balance} = \text{previous balance} - \text{principal reduction}$

Give all values to the nearest cent.

Payment number	Payment	Interest	Principal reduction	Balance
0	0.00	0.00	0.00	310 000.00
1	9220.00	7068.00	2152.00	307 848.00
2	9220.00	7018.93		
3				
4				

MATCHED EXAMPLE 7 | Analysing reducing balance loan amortisation tables

Monica has taken out a short-term business loan of $115 000. The interest rate is 20% p.a. compounding quarterly and it needs to be repaid by making four quarterly payments of $32 431. Answer the following questions using the amortisation table of the loan given below which has some missing entries.

Payment number	Payment	Interest	Principal reduction	Balance
0	0.00	0.00	0.00	115 000.00
1	32 431.00	5750.00	26 681.00	88 319.00
2	32 431.00	4415.95	28 015.05	60 303.95
3	32 431.00		29 415.80	
4	32 431.00	1544.41		1.55

Steps	Working

a What is the balance after three payments?

balance = previous balance −
principal reduction

b How much has the principal been reduced by at payment number 4?

principle reduction = payment −
interest

c How much interest was paid at payment 3?

1 Find r, the percentage interest rate
per compounding period.

2 interest = $\dfrac{r}{100}$ × previous balance

d How much of the principal has been paid after three payments?

Add the values in the principal
reduction column up to payment
number 3.

e How much has the loan cost Monica?

Add all the values in the interest
column.

f Explain how you know the loan isn't paid out fully after the four payments.

Look at the balance after the last
payment.

g If Monica wants to fully pay the loan with payment 4, what should the fourth payment be adjusted to?

Add the leftover amount to the last
payment.

SB

p. 350

a Anna has taken out a loan with a recurrence relation for V_n, the value of the loan after n compounding periods, of

$V_0 = 20\,250$, $V_{n+1} = 1.012V_n - 243$

Find V_1 and V_2 and explain why this shows that this model is an interest-only loan.

b Tiana has taken out an interest-only loan with an interest rate of 12% per annum compounding quarterly with regular quarterly payments of $3000. How much did Tiana borrow?

c Ariel has taken out a $115\,000 interest-only loan. The interest rate is 2.4% per annum compounding monthly. What are Ariel's regular monthly repayments?

Steps	Working
a 1 Find V_1.	
2 Find V_2.	
3 For interest-only loans, $V_n = V_0$.	
b 1 Identify what we know and what we need to find from the interest-only loan formula.	
2 Substitute into the formula and solve, using CAS if necessary.	
3 Write the answer.	
c 1 Identify what we know and what we need to find from the interest-only loan formula.	
2 Substitute into the formula and solve.	
3 Write the answer.	

MATCHED EXAMPLE 9 | Using finance solvers for annual loan interest rates

Patrick's reducing balance loan of $110 000 is to be fully repaid over 18 years with monthly repayments of $1008.16. Find the interest rate per annum correct to one decimal place and explain why you need to enter the payment as a negative.

Steps	Working
	TI-Nspire + **ClassPad**
1 Find I, given there are $18 \times 12 = 216$ months.	
2 Total number of compounding periods	N
Annual interest rate	I%
Present value for a loan is positive.	PV
Money moving away from a person is negative.	Pmt or PMT
Future value is zero when a loan is fully repaid.	FV
Same as CpY or C/Y	PpY or P/Y
Number of compounding periods per year	CpY or C/Y
3 After entering all the other values, the I% field displays the annual interest rate. Round to the correct number of decimal places.	
4 Is the money moving to the person or away from the person?	

MATCHED EXAMPLE 10	Using finance solvers for loan repayments, total loan cost and total interest paid

Gimli borrowed $40 000 at 6.8% per annum for eight years with quarterly repayments and interest compounding quarterly. Find, to the nearest cent

Steps	**Working**

a the repayment required to repay the loan in full.

1 Find Pmt, given there are 8 × 4 = 32 quarters.	
2 Total number of compounding periods	N
Annual interest rate	I%
Present value for a loan is positive.	PV
Money moving away from a person is negative.	Pmt or PMT
Future value is zero when a loan is fully repaid.	FV
Same as CpY or C/Y	PpY or P/Y
Number of compounding periods per year	CpY or C/Y
3 Write the answer correct to the nearest cent.	

b the total cost of the loan.

Total loan cost = N × Pmt

Ignore the negative sign in the Pmt value.

Round to the nearest cent.

c the total interest paid on the loan.

Total interest paid = N × Pmt − (PV − FV)

Ignore the negative signs in values.

Round to the nearest cent.

MATCHED EXAMPLE 11 Using finance solvers for amount owed and number of loan repayments

James takes out a loan of $155 000 at a rate of 7.2% per annum, compounding quarterly, with repayments of $4000.00 per quarter.

Steps	Working

a How much is still owing on James' loan after three years to the nearest cent?

 TI-Nspire **+** **ClassPad**

1 Find FV after $3 \times 4 = 12$ quarters.	
2 Total number of compounding periods	N
Annual interest rate	I%
Present value for a loan is positive.	PV
Money moving away from a person is negative.	Pmt or PMT
Future value has the opposite sign to present value.	FV
Same as CpY or C/Y	PpY or P/Y
Number of compounding periods per year	CpY or C/Y
3 Round to the nearest cent.	

b By what percentage has James reduced the balance of his loan? Round to the nearest percentage.

Percentage decrease in loan balance =

$$\frac{PV - FV}{PV} \times 100\%$$

Ignore the negative signs in the values.

Round to the nearest percentage.

c How many repayments are required in order for him to repay the loan in full?

1 Find N when FV = 0.	
2 Total number of compounding periods	N
Annual interest rate	I%
Present value for a loan is positive.	PV
Money moving away from a person is negative.	Pmt or PMT
Future value is zero when a loan is fully repaid.	FV
Same as CpY or C/Y	PpY or P/Y
Number of compounding periods per year	CpY or C/Y
3 When solving for N, always round *up*, never down, to the nearest whole number.	

d How much is still owing on James' loan after 67 repayments to the nearest cent?

1 Find FV after 67 quarters.	
2 Total number of compounding periods	N
Annual interest rate	I%
Present value for a loan is positive.	PV
Money moving away from a person is negative.	Pmt or PMT
Future value has the opposite sign to present value.	FV
Same as CpY or C/Y	PpY or P/Y
Number of compounding periods per year	CpY or C/Y

3 After entering all the other values, the FV field displays the amount still owing. Round to the nearest cent.

e James decides to adjust the value of the 67th repayment so that the loan is fully repaid. What will the adjusted 67th payment be to the nearest cent?

For the last payment, add the amount still owing on the loan to the regular payment, rounding to the nearest cent.

MATCHED EXAMPLE 12 | Using finance solvers for interest-only loans

Use finance solvers to find each of the following.

Steps	Working
a Lucas has a $60 000 interest-only loan compounding quarterly with quarterly payments of $930. What is the annual interest rate correct to one decimal place?	

TI-Nspire + **ClassPad**

1 Find I when the value of the loan stays the same.	
2 All compounding periods are the same for interest-only loans.	N
Annual interest rate	I%
Present value for a loan is positive.	PV
Money moving away from a person is negative.	Pmt or PMT
FV = −PV for interest-only loans	FV
Same as CpY or C/Y	PpY or P/Y
Number of compounding periods per year	CpY or C/Y
Round your answer to one decimal place.	

b Mia has taken out a $25 000 interest-only loan. The interest rate is 5.8% per annum compounding monthly. What are Mia's regular monthly repayments to the nearest cent?	

1 Find Pmt when the value of the loan stays the same.	
2 All compounding periods are the same for interest-only loans.	N
Annual interest rate	I%
Present value for a loan is positive.	PV
Money moving away from a person is negative.	Pmt or PMT
FV = −PV for interest-only loans	FV
Same as CpY or C/Y	PpY or P/Y
Number of compounding periods per year	CpY or C/Y
3 Round your answer to the nearest cent.	

Henry wants to buy a car and has taken out a car loan of $90 000 at an interest rate of 3.86% per annum, fixed for five years. Interest is calculated quarterly, and quarterly repayments are set at $1980. After five years, Henry renegotiates the conditions for the balance of his loan. The new interest rate will be 3.05% per annum. He will pay $2200 per quarter. How many years will it take him to repay the loan in full?

Steps	Working
	TI-Nspire + **ClassPad**
1 Find FV after $5 \times 4 = 20$ quarters for the first part of the loan.	
2 Total number of compounding periods	N
Annual interest rate for the first part of the loan	I%
Present value for a loan is positive.	PV
Money moving away from a person is negative.	Pmt or PMT
Future value has the opposite sign to present value.	FV
Same as CpY or C/Y	PpY or P/Y
Number of compounding periods per year	CpY or C/Y
3 Use the unrounded FV value as the positive PV value to find N for the second part of the loan.	
4 Total number of compounding periods	N
Annual interest rate for the second part of the loan	I%
Present value for a loan is positive.	PV
Money moving away from a person is negative.	Pmt or PMT
Future value is zero when a loan is fully repaid.	FV
Same as CpY or C/Y	PpY or P/Y
Number of compounding periods per year.	CpY or C/Y
5 Always round up your answer for N.	
Convert to years.	
6 Add the lengths of the two parts of the loan.	

MATCHED EXAMPLE 14 | Changing the interest rate

Sam borrows $200 500 at 6% per annum compounding monthly to be repaid over 12 years. The repayments for the first six years are $1995.25 each month. After six years, the interest rate is reduced to 5.2% per annum.

Steps	Working

a What is the new repayment to the nearest cent required for Sam to repay the loan?

TI-Nspire **+** **ClassPad**

1 Find FV after $6 \times 12 = 72$ months for the first part of the loan.

2 Total number of compounding periods   N

  Annual interest rate   I%

  Present value for a loan is positive.   PV

  Money moving away from a person is negative.   Pmt or PMT

  Future value has the opposite sign to present value.   FV

  Same as CpY or C/Y   PpY or P/Y

  Number of compounding periods per year   CpY or C/Y

3 Use the unrounded FV as the positive PV for the second part of the loan and find Pmt.

4 Total number of compounding periods   N

  Annual interest rate   I%

  Present value for a loan is positive.   PV

  Money moving away from a person is negative.   Pmt or PMT

  Future value is zero when a loan is fully repaid.   FV

  Same as CpY or C/Y   PpY or P/Y

  Number of compounding periods per year   CpY or C/Y

5 Write your answer to the nearest cent.

b How much total interest to the nearest cent has Sam paid over the 12 years?

1 Use the formula for both parts of the loan:

Total interest paid $= N \times Pmt - (PV - FV)$

Ignore the negative sign in values.

2 Add the two totals.

MATCHED EXAMPLE 15	Changing the repayments

Rachel has taken out a personal loan of $35 000 at the rate of 6% per annum, compounding monthly and with regular monthly payments, to pay for her wedding.

Steps	Working

a Rachel will make interest-only repayments for the first three years of this loan. How much is each interest-only repayment to the nearest dollar?

TI-Nspire + ClassPad

1 Find Pmt when the value of the loan stays the same.	
2 All compounding periods are the same for interest-only loans.	N
Annual interest rate	I%
Present value for a loan is positive	PV
Money moving away from a person is negative.	Pmt or PMT
$FV = -PV$ for interest-only loans	FV
Same as CpY or C/Y	PpY or P/Y
Number of compounding periods per year	CpY or C/Y
3 Write your answer to the nearest dollar.	

b For the next two years, Rachel will increase her monthly repayments so that the balance of the loan is $5250. What are Rachel's repayments, to the nearest cent, each month during these three years?

1 Find Pmt for the next $3 \times 12 = 36$ months.	
2 Total number of compounding periods	N
Annual interest rate	I%
Present value for a loan is positive.	PV
Money moving away from a person is negative.	Pmt or PMT
Future value has the opposite sign to present value.	FV
Same as CpY or C/Y	PpY or P/Y
Number of compounding periods per year	CpY or C/Y
3 Write your answer to the nearest cent.	

c Rachel will fully repay the outstanding balance of $5250 over the next two years. The first 23 monthly repayments will each be $280. The 24th repayment will have a different value to ensure the loan is repaid exactly to the nearest cent. What is the value of the 24th repayment, rounded to the nearest cent?

1 Find FV after the next $2 \times 12 = 24$ months of the loan.	
2 Total number of compounding periods	N
Annual interest rate	I%
Present value for a loan is positive.	PV
Money moving away from a person is negative.	Pmt or PMT
Future value has the opposite sign to present value.	FV
Same as CpY or C/Y	PpY or P/Y
Number of compounding periods per year	CpY or C/Y
3 For the last payment, add the amount still owing on the loan to the regular payment, rounding to the nearest cent.	

MATCHED EXAMPLE 16 | Working with annuity recurrence relations

William plans to invest $135 000 in an annuity at a rate of 10.2% per annum compounding monthly where he makes regular monthly withdrawals of $150.

a Write a recurrence relation for the balance.

William changes his mind and decides to invest the $135 000 in an annuity compounding weekly with weekly withdrawals that has the recurrence relation $V_0 = 135 000$, $V_{n+1} = 1.003V_n - 150$

b What is the annual interest rate of this investment?

c How much has the investment increased between the third- and fourth-week period, to the nearest cent?

Steps	Working
a 1 Find the number of compounding periods per year.	
2 Identify V_n, V_0, r and d.	
3 Substitute the values into these formulas and simplify. V_0 = principal $V_{n+1} = \left(1 + \dfrac{r}{100}\right)V_n - d$	
b 1 Use the recurrence relation to find an equation for r, the percentage interest rate per compounding period.	
2 Solve for r, using CAS if necessary.	
3 Multiply r by the compounding period to find the annual percentage compound interest rate.	

c Use CAS recursive computation to find the balances, and subtract to find the difference, rounding to the nearest cent.

TI-Nspire

ClassPad

6

MATCHED EXAMPLE 17 | Analysing annuity amortisation tables

The following amortisation table shows the first three payments of an annuity with quarterly compounding interest and quarterly withdrawals. Some of the entries are missing.

Payment number	Payment	Interest	Principal reduction	Balance
0	0.00	0.00	0.00	500 000.00
1	10 250.00	9000.00	1250.00	498 750.00
2	10 250.00			
3	10 250.00	8954.60	1295.41	496 182.10

Find the following:

Steps	Working
a the amount invested	
Read the principal from the table.	
b the regular quarterly withdrawal	
Read the regular payment from the table.	
c the interest rate per compounding period	
$r = \dfrac{\text{interest}}{\text{previous balance}} \times 100$ Choose the option with the easiest calculations.	
d the interest rate per annum	
Multiply r by the compounding period.	
e the interest paid at payment 2	
$\text{interest} = \dfrac{r}{100} \times \text{previous balance}$	
f the amount by which the principal has been reduced at payment 2	
principal reduction = payment − interest	
g the balance after two withdrawals	
balance = previous balance − principal reduction	

9780170464055

MATCHED EXAMPLE 18 | Calculating how long an annuity will last

Michelle purchases a $550 000 annuity. Interest is paid at 7.2% per annum compounding monthly. If she receives monthly payments of $4950, how many years will the annuity last?

Steps	Working
	TI-Nspire + **ClassPad**
1 Find N when FV = 0.	
2 Total number of compounding periods	N
Annual interest rate	I%
Present value for an investment is negative.	PV
Money moving to a person is positive.	Pmt or PMT
Future value is zero when an investment is fully paid out.	FV
Same as CpY or C/Y	PpY or P/Y
Number of compounding periods per year	CpY or C/Y
3 Always round N *up* to the next whole number.	
4 Convert to years and answer the question.	

6

MATCHED EXAMPLE 19 | Calculating how much to withdraw from an annuity

Maria invests $400 000 in an annuity with interest paid at 10.2% per annum compounding monthly. She receives monthly payments from this investment. What monthly payment will she receive, to the nearest cent, if she wishes to receive payments for 18 years?

Steps	Working
	TI-Nspire + **ClassPad**
1 Find Pmt, given there are $18 \times 12 = 216$ months.	
2 Total number of compounding periods	N
Annual interest rate	I%
Present value for an investment is negative.	PV
Money moving to a person is positive.	Pmt or PMT
Future value is zero when an investment is fully paid out.	FV
Same as CpY or C/Y	PpY or P/Y
Number of compounding periods per year	CpY or C/Y
3 Round your answer to the nearest cent.	

MATCHED EXAMPLE 20 | Working with perpetuities

a Annette has an investment whose recurrence relation for V_n, the value of the investment after n compounding periods, is

$V_0 = 75\,000, V_{n+1} = 1.008V_n - 600$

Find V_1 and V_2 and explain why this shows that this is a perpetuity.

b Michael wishes to set up a scholarship fund in his name so that each year an amount of $2400 is awarded to a young filmmaker at the school he attended. If interest on the investment is 6% per year, compounded annually, how much should he invest in this perpetuity?

c Rebecca has $450\,000 to set up a perpetuity for her nephew Peter. She invests the money in bonds that return 2.8% per annum compounding quarterly. Use the perpetuity formula to find how much Peter will receive each quarter from this investment, rounding to the nearest cent.

Steps	Working
a 1 Find V_1.	
2 Find V_2.	
3 For perpetuities $V_n = V_0$.	
b 1 Identify what we know and what we need to find from the perpetuity formula.	
2 Substitute into the formula and solve, using CAS if necessary.	
3 Write the answer.	
c 1 Identify what we know and what we need to find from the perpetuity formula.	
2 Substitute into the formula and solve.	
3 Round your answer to the nearest cent.	

MATCHED EXAMPLE 21 Using finance solvers for perpetuities

Use finance solvers to find each of the following.

Steps	Working

a An art school invests $150 000 in a perpetuity. The interest earned from this perpetuity will provide an annual art prize of $8000 for the most creative student. What annual interest rate, correct to one decimal place, would be required for this investment?

TI-Nspire **+** **ClassPad**

1 Find I when the value of the investment stays the same.

2 All compounding periods are the same for perpetuities.   N

Annual interest rate   I%

Present value for an investment is negative.   PV

Money moving to a person is positive.   Pmt or PMT

FV = −PV for perpetuities   FV

Same as CpY or C/Y   PpY or P/Y

Number of compounding periods per year   CpY or C/Y

3 Round your answer to one decimal place.

b Rebecca has $450 000 to set up a perpetuity for her nephew Peter. She invests the money in bonds that return 2.8% per annum compounding quarterly. Use a finance solver to find how much Peter will receive each quarter from this investment, rounding to the nearest cent.

1 Find Pmt when the value of the investment stays the same.

2 All compounding periods are the same for perpetuities.   N

Annual interest rate   I%

Present value for an investment is negative.   PV

Money moving to a person is positive.   Pmt or PMT

FV = −PV for perpetuities   FV

Same as CpY or C/Y   PpY or P/Y

Number of compounding periods per year   CpY or C/Y

3 Round your answer to the nearest cent.

MATCHED EXAMPLE 22 | Working with annuity investment recurrence relations

James plans to invest $85 000 in an annuity investment a rate of 5.2% per annum compounding quarterly where he makes regular quarterly additions of $550.

a Write a recurrence relation for the balance.

James then changes his mind and decides to invest the $85 000 in a different annuity investment compounding monthly with monthly additions that has the recurrence relation

$$V_0 = 85\,000, \; V_{n+1} = 1.0035V_n + 550$$

b What is the annual interest rate of this investment?

c How much has the investment increased between the third and fourth month, to the nearest cent?

Steps	Working
a 1 Find the number of compounding periods per year.	
2 Identify V_n, V_0, r and d.	
3 Substitute the values into $V_0 = $ principal, $V_{n+1} = \left(1 + \dfrac{r}{100}\right)V_n + d$ and simplify.	
b 1 Use the recurrence relation to find an equation for r, the percentage interest rate per compounding period.	
2 Solve for r, using CAS if necessary.	
3 Multiply r by the compounding period to find the annual percentage compound interest rate.	

c Use CAS recursive computation to find the balances, and subtract to find the difference, rounding to the nearest cent.

TI-Nspire

ClassPad

9780170464055

SB

p. 385

Carrie invests $6200 in an annuity investment earning interest compounding annually. She deposits an extra $600 into the account each year after the initial deposit. The amortisation table below shows the first few calculations for her investment.

Payment number	Payment	Interest	Principal addition	Balance
0	0.00	0.00	0.00	6200.00
1	600.00	$\frac{r}{100} \times 6200 = 620.00$	$600.00 + 620.00 = 1220.00$	$6200.00 + 1220.00 = 7420.00$
2	600.00	$\frac{r}{100} \times 7420 = 742.00$		
3	600.00			
4				

Steps | **Working**

a Use the table to show two calculations that will give r, the percentage interest rate per compounding period.

Use $r = \dfrac{\text{interest}}{\text{previous balance}} \times 100$

b What is the nominal interest rate for the loan?

Use the compounding period to find the annual interest rate.

c Complete the amortisation table, showing the interest paid, principal reduction and balance for the first four years of the loan, giving all values to the nearest cent.

Complete the table using

interest $= \dfrac{r}{100} \times$ previous balance

principal addition = payment + interest

balance = previous balance + principal addition

Give all values to the nearest cent.

Payment number	Payment	Interest	Principal addition	Balance
0	0.00	0.00	0.00	6200.00
1	600.00	620.00	1220.00	7420.00
2	600.00	742.00		
3	600.00			
4				

MATCHED EXAMPLE 24 | Analysing annuity investment amortisation tables

Jordan has invested $8500.00 in an annuity investment with an interest rate of 8% p.a. compounding quarterly, and he is making additional quarterly payments of $800. Answer these questions using the amortisation table of the investment given below which has some missing entries, giving your answers to the nearest cent where relevant.

Payment number	Payment	Interest	Principal addition	Balance
0	0	0	0	8500
1	800.00	170.00	970.00	9470.00
2	800.00			10 459.40
3	800.00	209.19	1009.19	11 468.59
4		229.37	1109.37	12 577.96

Steps	Working
a Find r, the interest rate per compounding period.	
Divide the percentage interest rate per year by the number of compounding periods per year.	
b What is the missing interest earned in the second quarter?	
interest = $\dfrac{r}{100}$ × previous balance	
c What is the missing principal addition in the second quarter?	
principal addition = payment + interest	
d Jordan paid a higher amount in the fourth quarter. How much is the payment?	
principal addition = payment + interest	

9780170464055

MATCHED EXAMPLE 25	Using finance solvers for annuity investments

Alex's investment earns interest at the rate of 6.2% per annum, compounding quarterly. Alex initially invested $70 000 and adds monthly payments of $850.

SB

p. 387

Steps	Working

a After how many quarters will the value of this investment first exceed $85 000?

	TI-Nspire + **ClassPad**
1 Enter the amount as FV and find N.	
2 Total number of compounding periods	N
Annual interest rate	I%
Present value for an investment is negative.	PV
Money moving away from a person is negative.	Pmt or PMT
Future value has the opposite sign to present value.	FV
Same as CpY or C/Y	PpY or P/Y
Number of compounding periods per year	CpY or C/Y
3 Round N *up* to the nearest whole number.	

b Alex wants to reach a target of $115 000 in five years. After two years, Alex increased his payments so that he would reach his target. What did Alex increase his payments to? Give your answer to the nearest cent.

1 Find FV after eight quarters.	
2 Total number of compounding periods	N
Annual interest rate	I%
Present value for an investment is negative.	PV
Money moving away from a person is negative.	Pmt or PMT
Future value has the opposite sign to present value.	FV
Same as CpY or C/Y	PpY or P/Y
Number of compounding periods per year	CpY or C/Y
3 Use FV after one year as PV for the next three years.	
4 Total number of compounding periods	N
Annual interest rate	I%
Present value for an investment is negative.	PV
Money moving away from a person is negative.	Pmt or PMT
Future value has the opposite sign to present value.	FV
Same as CpY or C/Y	PpY or P/Y
Number of compounding periods per year	CpY or C/Y
5 Write your answer to the nearest cent.	

MATCHED EXAMPLE 26 Using finance solvers for different types of compound interest investments

Addison started her savings by investing $12 500 for eight years in an account earning 3.2% p.a. interest compounding quarterly. After the eight years, she put the balance into a superannuation account for her retirement earning 7.4% p.a. interest compounding quarterly and made regular quarterly payments of $650 for 34 years until she retired. Since Addison retired, she has been taking out regular quarterly payments of $3800 from this account. How much money, to the nearest cent, is in her account after she has been retired for eight years?

Steps	Working
	TI-Nspire + **ClassPad**
The first account is an ordinary compound interest investment. Find FV after 8 × 4 = 32 quarters.	
1 Total number of compounding periods	N
Annual interest rate	I%
Present value for an investment is negative.	PV
Payment amount	Pmt or PMT
Future value has the opposite sign to present value.	FV
Same as CpY or C/Y	PpY or P/Y
Number of compounding periods per year	CpY or C/Y
2 The superannuation account is an annuity investment. Enter the unrounded FV from the first account as the new PV. Find FV after 34 × 4 = 136 quarters.	
3 Total number of compounding periods	N
Annual interest rate	I%
Present value for an investment is negative.	PV
Money moving away from a person is negative.	Pmt or PMT
Future value has the opposite sign to present value.	FV
Same as CpY or C/Y	PpY or P/Y
Number of compounding periods per year	CpY or C/Y
The superannuation account is now an annuity. Enter the unrounded FV from the annuity account as the new PV. Find FV after 8 × 4 = 32 quarters.	
4 Total number of compounding periods	N
Annual interest rate	I%
Present value for an investment is negative.	PV
Money moving to a person is positive.	Pmt or PMT
Future value has the opposite sign to present value.	FV
Same as CpY or C/Y	PpY or P/Y
Number of compounding periods per year	CpY or C/Y
5 Write your answer to the nearest cent.	

MATRICES AND THEIR APPLICATIONS

CHAPTER

7

SB

p. 407

MATCHED EXAMPLE 1 | Understanding the order of matrices

For the table showing the information about the number of different lollies sold at different places, find the following.

	Supermarket	Theatre	Petrol station	Pharmacy
Sherbies	50	20	45	20
Pods	100	50	35	10
Fads	120	60	65	15
Cheekies	95	35	70	30

Steps	Working

a The matrix *A* that could be used to show this information, stating its order and number of elements.

Rewrite the information in the table as a matrix.

b The matrix that could be used to show the number of pods sold at the petrol station and state its order.

Find the information in the table and write as a matrix.

c The 1 × 4 matrix that could be used to show the number of fads sold in each of the places.

Find the information in the table and write as a matrix.

d The 1 × 4 matrix that could be used to show the number of each type of lolly sold in the supermarket.

Find the information in the table and write as a matrix.

---

e The 2×2 matrix that could be used to show the number of sherbies and fads sold at the supermarket and petrol station.

Find the information in the table and write as a matrix.

f The 4×1 matrix that could be used to show the total for each of the four types of lollies sold.

Find the information in the table and write as a matrix.

g Copy the following labelled matrix showing the information from the table and fill in the missing numbers. The lolly types are shown by S = Sherbies, P = Pods, F = Fads and C = Cheekies

$$
\begin{array}{c}
\\
\text{Supermarket} \\
\text{Theatre} \\
\text{Petrol station} \\
\text{Pharmacy}
\end{array}
\begin{array}{cccc}
S & P & F & C \\
\left[\begin{array}{cccc}
& & 120 & \\
& & & 35 \\
& & & \\
& 10 & &
\end{array}\right]
\end{array}
$$

Find the information in the table and complete the matrix.

MATCHED EXAMPLE 2 | Identifying types of matrices

a For each of the following matrices, state the order and whether it is a row, column, summing, square, zero, binary or permutation matrix.

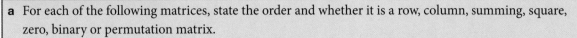

a $\begin{bmatrix} 1 & 1 & 0 \\ 0 & 0 & 0 \\ 0 & 1 & 1 \end{bmatrix}$

b $\begin{bmatrix} 0 & 0 & 0 & 0 \\ 0 & 0 & 0 & 0 \\ 0 & 0 & 0 & 0 \\ 0 & 0 & 0 & 0 \end{bmatrix}$

c $\begin{bmatrix} 1 & 0 & 0 & 0 \\ 0 & 0 & 0 & 1 \\ 1 & 0 & 0 & 0 \\ 0 & 0 & 0 & 1 \end{bmatrix}$

d $\begin{bmatrix} 1 \\ 1 \\ 1 \\ 1 \\ 1 \end{bmatrix}$

e $\begin{bmatrix} 0 & 3 & 0 \\ 4 & 0 & 0 \\ 0 & 0 & 3 \end{bmatrix}$

Steps	Working
Does the matrix have	

Does the matrix have

- just one row or just one column
- the same number of rows and columns
- all zeros
- just zeros and ones
- exactly one '1' in every row and column and zeros everywhere else?

SB

p. 410

MATCHED EXAMPLE 3	Finding the transpose of matrices

Find the transpose of the following matrices and state the orders of both A and A^T.

a $A = \begin{bmatrix} 4 & 3 & 1 \\ 0 & 0 & 1 \\ 2 & -1 & 2 \\ 3 & 2 & 4 \end{bmatrix}$

b $A = \begin{bmatrix} 1 & 1 \\ 2 & 3 \\ 4 & 0 \end{bmatrix}$

c $A = \begin{bmatrix} 1 \\ 2 \\ 0 \\ 3 \end{bmatrix}$

Steps	Working
a Switch the rows and columns of A.	
b Switch the rows and columns of A.	
c Switch the rows and columns of A.	

SB

Using CAS 1:
Working with
matrices
p. 410

9780170464055

MATCHED EXAMPLE 4 | Identifying types of square matrices using leading diagonals

For each of the following matrices, list the numbers in the leading diagonal and state whether it is a symmetric, upper triangular, lower triangular, diagonal matrix or an identity matrix. If the matrix is not any of these, explain why.

a $\begin{bmatrix} 1 & 2 & 1 & 0 \\ -1 & 2 & 0 & 0 \\ 2 & 0 & 0 & 1 \end{bmatrix}$

b $\begin{bmatrix} 1 & 1 & 2 & 3 \\ 0 & 2 & 2 & -1 \\ 0 & 0 & 3 & 2 \\ 0 & 0 & 0 & 0 \end{bmatrix}$

c $\begin{bmatrix} 1 & 2 & -3 & 3 & 2 \\ 2 & 0 & 4 & -5 & 1 \\ -3 & 4 & 3 & -4 & 3 \\ 3 & -5 & -4 & 2 & 2 \\ 2 & 1 & 3 & 2 & -1 \end{bmatrix}$

d $\begin{bmatrix} 1 & 0 & 0 \\ 0 & 1 & 0 \\ 2 & 2 & 1 \end{bmatrix}$

e $\begin{bmatrix} 1 & 1 & 0 \\ 0 & 1 & 1 \\ 0 & 0 & 1 \end{bmatrix}$

Steps	Working
Is it a square matrix?	
Is the matrix the same as its transpose?	
Does the matrix have	
• all zero elements except the ones in the leading diagonal	
• a leading diagonal made up of only ones	
• all zero elements below the leading diagonal	
• all zero elements above the leading diagonal?	

MATCHED EXAMPLE 5 | Using element notation

A fruit stall owner keeps track of the number of types of fruits sold on different days of the week. The matrix M gives the number of oranges (O), nectarines (N), plums (P), and apricots (A) sold on four days of the week. The element in row i and column j of matrix M is m_{ij}.

$$M = \begin{array}{c} \\ Mon \\ Tue \\ Wed \\ Thurs \end{array} \begin{array}{cccc} O & N & P & A \\ \left[\begin{array}{cccc} 20 & 10 & 12 & 20 \\ 15 & 12 & 14 & 18 \\ 17 & 15 & 10 & 16 \\ 11 & 16 & 12 & 15 \end{array} \right] \end{array}$$

What information about the stock does each the following give?

a m_{22}

b m_{34}

c $m_{11} + m_{21} + m_{31} + m_{41}$

d $m_{41} + m_{42} + m_{43} + m_{44}$

Steps	Working
a m_{22} is the element in the second row second column.	
b m_{34} is the element in the third row fourth column.	
c $m_{11} + m_{21} + m_{31} + m_{41}$ is the sum of all the elements in the first column.	
d $m_{41} + m_{42} + m_{43} + m_{44}$ is the sum of all the elements in the fourth row.	

MATCHED EXAMPLE 6 | Working with matrix elements

Construct the matrix A for each set of rules about the elements a_{ij}, where i is the row number and j is the column number.

Steps	Working
a A is a 3×3 matrix, where $a_{ij} = 1$ when $i = j$, $a_{12} = 2$, $a_{13} = 3$, $a_{23} = 0$ and $a_{ij} = -a_{ji}$ when $i \neq j$.	
1 List the elements of the matrix in a_{ij} form.	
2 Use the rule to find all the elements of A.	
b A is the transpose of B. B is a 2×4 matrix, where $b_{ij} = j$.	
1 List the elements of matrix B in b_{ij} form.	
2 Use the rule to find all the elements of B.	
3 Find A, the transpose of B.	
c A is a 2×2 matrix, where $a_{ij} = i + j + 3$ when $i = j$ and all the other elements are 0.	
1 List the elements of the matrix in a_{ij} form.	
2 Use the rule to find all the elements of A.	

d A is a 3×2 matrix, where $a_{ij} = 0$ when $i < j$ and $a_{ij} = 1$ otherwise.

1 List the elements of the matrix in a_{ij} form.

2 Use the rule to find all the elements of A.

MATCHED EXAMPLE 7 | Adding and subtracting matrices, and multiplying matrices by a scalar

If $A = \begin{bmatrix} 1 & 0 & 2 \end{bmatrix}$, $B = \begin{bmatrix} 1 & 2 \\ 0 & -1 \\ 0 & 0 \end{bmatrix}$, $C = \begin{bmatrix} 3 & 1 & 2 \end{bmatrix}$ and $D = \begin{bmatrix} 2 & 0 \\ 0 & 1 \\ 4 & 2 \end{bmatrix}$, calculate the following, giving a

reason if the addition or subtraction is not defined.

a $4B$

b $A - C$

c $B + D$

d $3C - B$

e $A - \dfrac{1}{2} C$

f $0.25B$

Steps	Working
1 Check that the matrices have the same order.	
2 Add or subtract the corresponding elements.	
3 Multiply each element by the scalar.	

SB

p. 424

MATCHED EXAMPLE 8 | Solving matrices using addition, subtraction and scalar multiplication

Solve each of the following.

Steps	Working

a Find x and y if $\begin{bmatrix} x & -1 \\ -1 & 0 \\ 3 & 2 \end{bmatrix} - \dfrac{1}{2} \times \begin{bmatrix} 0 & -1 \\ 1 & 3 \\ y & 2 \end{bmatrix} = \begin{bmatrix} 12 & -\dfrac{1}{2} \\ -\dfrac{3}{2} & -\dfrac{3}{2} \\ 2 & 0 \end{bmatrix}$

1 Simplify by doing the matrix addition, subtraction and scalar multiplication until there is one matrix on either side of the equal sign.

2 Use the fact that two equal matrices must have all the same elements in the same places and solve, using CAS if necessary.

b Find A if $\begin{bmatrix} 2 & -2 \\ 0 & 0 \end{bmatrix} + A = 2 \times \begin{bmatrix} 2 & 5 \\ 4 & 2 \end{bmatrix} - \begin{bmatrix} 3 & 0 \\ 3 & 0 \end{bmatrix}$

1 Simplify by doing the matrix addition, subtraction and scalar multiplication.

2 Solve for the unknown matrix by using matrix addition, subtraction and scalar multiplication.

c *A* and *B* are both 3×3 matrices. *A* has elements that follow the rule $a_{ij} = 1$ when $i = j$, and $a_{ij} = 2$ when $i \neq j$. *B* has elements that follow the rule $b_{ij} = 3i + j - 2$. Find $2A+B$.

   1 Find *A* using the element rule.

   2 Find *B* using the element rule.

   3 Use matrix addition, subtraction and scalar multiplication.

MATCHED EXAMPLE 9 | Working with matrices using addition, subtraction and scalar multiplication

The ages of four people working in the same company are 48, 40, 42 and 44. Each person was half their age when they bought their first car.

Steps	**Working**
a Show a matrix calculation involving a column matrix that will give the age at which each person bought their first car.	
Use scalar multiplication.	
b If each person got their driving licence two years before they bought their first car, show how this can be included in the matrix to calculate the age at which they got their driving licences.	
Use matrix subtraction.	
c If each person got their first promotion 10 years after they got their driving licence, show how this can be included in the matrix to calculate the age at which they got their first promotion.	
Use matrix addition.	

SB

p. 430

If $M = \begin{bmatrix} 2 \\ 0 \\ 1 \end{bmatrix}$, $N = \begin{bmatrix} 4 & 1 \\ 1 & 0 \\ -2 & 3 \end{bmatrix}$, $O = \begin{bmatrix} 2 & 1 \\ 1 & 1 \end{bmatrix}$ and $P = \begin{bmatrix} 2 & -1 & 0 \end{bmatrix}$, for each of the following.

a MO

b NO

c MN

d MP

e N^4

f PM

g $O^2 - O$

    i state whether or not the expression is defined, giving a reason.

For those that are defined

    ii state the order of the answer before performing the calculation

    iii compute the matrix expressions to find the answer.

Steps	Working
a **i** Do the number of columns in M = number of rows in O?	

b **i** Do the number of columns in N = number of rows in O?
 	
   ii How many rows does N have?
 How many columns does O have?
 	
   iii Calculate NO.	

c **i** Do the number of columns in M = number of rows in N?	

d **i** Do the number of columns in $M =$ number of rows in P?

    ii How many rows does M have?

       How many columns does P have?

    iii Calculate MP.

e **i** Is N square matrix?

f **i** Do the number of columns in $P =$ number of rows in M?

       How many rows does P have?

       How many columns does M have?

       Calculate PM.

g **i** Is O square matrix?

    ii How many rows does O have?

       How many columns does O have?

       Is the calculation $O^2 - O$ possible?

    iii Calculate $O^2 - O$.

9780170464055

MATCHED EXAMPLE 11 | Working with matrix multiplication

Show each of the following.

Steps	Working

a Show that for the matrix equation $\begin{bmatrix} 1 & 2 & 0 \\ 1 & 1 & 0 \\ 2 & 0 & 1 \end{bmatrix} \begin{bmatrix} 3 \\ x \\ 1 \end{bmatrix} = \begin{bmatrix} 3 \\ y \\ 7 \end{bmatrix} + \begin{bmatrix} y \\ 1 \\ 0 \end{bmatrix}$ to be true, the equations $y = 2x$ and $y = 2 + x$ must be true.

7

   1 Multiply the matrices on the left-hand side of the equation and add the matrices on the right-hand side of the equation.

   2 Equate the two resulting matrices and use the fact that corresponding elements must be equal.

     Rearrange each of the equations to make y the subject.

b Show that if M is a square matrix, matrix N is a column matrix and matrix O is a row matrix, then MNO is a square matrix.

   1 Write the orders of each matrix.

   2 Use the fact that the product matrix has the same number of rows as the first matrix and the same number of columns as the second matrix.

SB

Using CAS 3:
Multiplication
and powers of
matrices
p. 433

MATCHED EXAMPLE 12 | Multiplying summing matrices

For the matrices $A = \begin{bmatrix} 2 & 3 \\ 5 & 1 \\ 2 & 3 \\ 1 & 0 \end{bmatrix}$ and $B = \begin{bmatrix} 2 & 6 & -2 \\ 4 & 3 & 2 \\ 1 & 0 & 3 \end{bmatrix}$, show how to use a summing matrix S to calculate

a the sums and means of the rows of B.

b the sums and means of the columns of A.

Steps	Working
a 1 Find m and n for B.	
2 What sort of summing matrix is needed?	
3 Do the multiplication to find the sums of each of the rows, using CAS if necessary.	
4 What do we need to multiply by to find the means of each of the rows?	
5 Perform the scalar multiplication to find the means of each of the rows, using CAS if necessary.	
b 1 Find m and n for A.	
2 What sort of summing matrix is needed?	
3 Do the multiplication to find the sums of each of the columns, using CAS if necessary.	

4 What do we need to multiply by to find the means of each of the columns?

5 Perform the scalar multiplication to find the means of each of the columns, using CAS if necessary.

7

MATCHED EXAMPLE 13 | Multiplying permutation matrices

Find the following matrices.

Steps	Working

a What is this matrix product equal to?

$$\begin{bmatrix} 0 & 0 & 1 & 0 & 0 & 0 \\ 0 & 1 & 0 & 0 & 0 & 0 \\ 1 & 0 & 0 & 0 & 0 & 0 \\ 0 & 0 & 0 & 0 & 1 & 0 \\ 0 & 0 & 0 & 0 & 0 & 1 \\ 0 & 0 & 0 & 1 & 0 & 0 \end{bmatrix} \begin{bmatrix} L \\ I \\ S \\ T \\ E \\ N \end{bmatrix}$$

1 Locate where the 1 appears in each row to identify where the letter has moved.

2 Complete the matrix product.

b Matrix P is a 5×5 permutation matrix and matrix M is another matrix such that the matrix product $M \times P$ is defined. This matrix product results in the entire first and fifth columns of matrix M being swapped. Find the permutation matrix P.

1 Locate where the 1 appears in each column.

2 Write the matrix.

MATCHED EXAMPLE 14 | Showing whether matrices are inverses of each other

Show that the matrices $\begin{bmatrix} 4 & 5 \\ 3 & 4 \end{bmatrix}$ and $\begin{bmatrix} 4 & -5 \\ -3 & 4 \end{bmatrix}$ are inverses of each other.

Steps	Working
1 Show that the product of the two matrices equals the identity matrix.	
2 **Reverse** the order of the two matrices and show that their product also equals the identity matrix.	

MATCHED EXAMPLE 15 | Finding the determinant and inverse of a matrix

For each of the following matrices find

   i the determinant

   ii the inverse (if it exists).

a $A = \begin{bmatrix} 3 & 6 \\ 1 & 2 \end{bmatrix}$ **b** $B = \begin{bmatrix} 8 & 12 \\ 1 & 2 \end{bmatrix}$ **c** $C = \begin{bmatrix} 3 & 3 \\ 4 & 3 \end{bmatrix}$

Steps	Working
a **i** For $A = \begin{bmatrix} a & b \\ c & d \end{bmatrix}$, use $\det(A) = ad - bc$.	
ii Find $A^{-1} = \dfrac{1}{\det(A)}\begin{bmatrix} d & -b \\ -c & a \end{bmatrix}$ (if it exists).	
b **i** For $B = \begin{bmatrix} a & b \\ c & d \end{bmatrix}$, use $\det(B) = ad - bc$.	
ii Find $B^{-1} = \dfrac{1}{\det(B)}\begin{bmatrix} d & -b \\ -c & a \end{bmatrix}$ (if it exists).	
c **i** For $C = \begin{bmatrix} a & b \\ c & d \end{bmatrix}$, use $\det(C) = ad - bc$.	
ii Find $C^{-1} = \dfrac{1}{\det(C)}\begin{bmatrix} d & -b \\ -c & a \end{bmatrix}$ (if it exists).	

SB

p. 442

a If the determinant of $\begin{bmatrix} a & -6 \\ 2 & 6 \end{bmatrix}$ is equal to 30, what is the value of a?

b For the matrix $A = \begin{bmatrix} 6 & 4 \\ 2 & a \end{bmatrix}$, what is the value of a for which $A = 2A^{-1}$?

c Show that if A and B are both square matrices and AB is defined, then $A^{-1}B^2 + B^{-1}A^2$ is defined.

7

Steps	Working
a **1** Find the determinant of the matrix.	
2 Let the determinant equal the number given and solve for a, using CAS if necessary.	
b **1** Find the determinant of A.	
2 Find A^{-1}.	
3 Let $A = 2A^{-1}$. Since every pair of corresponding elements has to be equal, set a pair of corresponding elements equal to each other. Solve for a, using CAS if necessary.	
c **1** State the orders of A and B.	
2 Use the information given about A and B.	
3 Use the orders of A and B to find the orders of the other matrices in the question.	
4 Write a matrix order equation and answer the question.	

SB

Using CAS 4:
Finding the
determinant and
inverse of a matrix
p. 443

p. 446

MATCHED EXAMPLE 17 | Solving sport problems using matrices

Each week, the coach of the Melbourne Little Champions basketball team awards the Best Player Award to the player who scores the most game points. The results of their last game were as follows.

	3 pointers	2 pointers	1 pointer
Kevin	1	1	0
Ron	0	1	1
Anna	0	0	0
Alex	2	0	2
Oliver	0	1	0
Andrew	0	0	2
Darcy	0	0	1

Steps	**Working**
a Write a column matrix, *P*, to represent the three different scores possible.	
When a player scores, they can either get 3 points, 2 points or 1 point.	
b Write a matrix, *M*, to represent the data from the table.	
Matrix *M* will have 7 rows and 3 columns.	
c Use matrix multiplication to find a score matrix, *S*, that represents the total scored by each of the players.	
To find *S*, we must find *MP*. This will give us a $(7 \times 3)(3 \times 1)$, which will result in a 7×1 matrix that represents the personal total score for each of the seven players.	
d Which player won the Best Player Award?	
The largest element in *S* is the highest personal score.	
e The opposition team scored a total of 28 points. Did the Melbourne Little Champions win the game?	
1 Find the total score for Melbourne Little Champions by adding all of the elements in *S*.	
2 Write the answer.	

9780170464055

MATCHED EXAMPLE 18 | Solving problems using summing matrices

Matrix T shows the number of T-shirts in the colours red (R), blue (B), green (G) and pink (P) made by members in a T-shirt designing class.

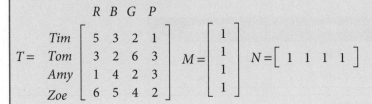

$$T = \begin{array}{c} \\ Tim \\ Tom \\ Amy \\ Zoe \end{array} \begin{array}{cccc} R & B & G & P \\ \left[\begin{array}{cccc} 5 & 3 & 2 & 1 \\ 3 & 2 & 6 & 3 \\ 1 & 4 & 2 & 3 \\ 6 & 5 & 4 & 2 \end{array} \right] \end{array} \quad M = \left[\begin{array}{c} 1 \\ 1 \\ 1 \\ 1 \end{array} \right] \quad N = \left[\begin{array}{cccc} 1 & 1 & 1 & 1 \end{array} \right]$$

Steps	Working
a **i** Calculate the product $P = TM$. **ii** What information does the element p_{31} give?	

a   **i**   Multiply the matrices. The matrix order equation tells us the order of the product:
 	
      $(4 \times 4)\,(4 \times 1) = 4 \times 1$
 	
      The product involves summing the rows of T.
 	
    ii p_{ij} is the element in the ith row and jth column in P.	

b **i** Calculate $Q = NT$ and describe the information given by Q. **ii** What information does the element q_{14} give?	

b   **i**   Multiply the matrices. The matrix order equation tells us the order of the product:
 	
      $(1 \times 4)(4 \times 4) = 1 \times 4$
 	
      The product involves summing the columns of T.
 	
    ii q_{ij} is the element in the ith row and jth column in Q.	

c Calculate $\dfrac{1}{4}\,NTM$ and describe the information given by this matrix.	

c   Perform the multiplication. The matrix order equation tells us the order of the product:
 	
     $(1 \times 4)\,(4 \times 1) = 1 \times 1$
 	
     The product involves adding elements and dividing by the number of elements.	

MATCHED EXAMPLE 19 | Solving costing and pricing problems using matrices

A sports equipment dealer purchases basketballs for $12 each and netballs for $10 each. In the last two months, he purchased the number of balls shown below.

	Basketballs	Netballs
Month 1	150	200
Month 2	105	115

Steps	Working

a Find the two matrices that can be multiplied to give the total purchase cost of balls in each of the two months and complete the multiplication.

We need a matrix product that calculates

no. of basketballs × cost of basketballs + no. of netballs × cost of netballs.

b The dealer sells goods at a 40% markup. He recorded his purchase costs over the last two months for balls and three other items in the following table.

	Month 1	Month 2
Cricket bats	$4500	$6000
Gloves	$2300	$1200
Balls		
Bags	$1000	$3250

i Represent these costs in a 4 × 2 cost matrix, C.

ii Using scalar multiplication, represent the selling prices of these goods in a 4 × 2 matrix, S.

i The table already has 4 rows and 2 columns. Fill in the missing information from part **a**.

ii 1 State what the markup means for the cost price in terms of the selling price.

2 Write an equation connecting the selling price and cost price by converting the percentage to a decimal.

3 Show this as a matrix equation and give the answer using scalar multiplication.

c   i   Create a profit matrix.

   ii   Calculate the total profit to be made if all the goods purchased over these two months are sold.

i   To create a profit matrix:

profit = selling price − cost price

ii   The total profit can be found by adding all of the elements in the profit matrix.

MATCHED EXAMPLE 20 | Working with communication matrices and diagrams

Steps	Working

a The communication matrix M shows how direct messages can be sent between four people: Zoe (Z), Chloe (C), Emma (E) and Alexis (A).

$$M = sender \begin{array}{c} \\ Z \\ C \\ E \\ A \end{array} \overset{\begin{array}{cccc} receiver \\ Z & C & E & A \end{array}}{\begin{bmatrix} 0 & 0 & 1 & 1 \\ 1 & 0 & 0 & 1 \\ 0 & 1 & 0 & 1 \\ 0 & 0 & 1 & 0 \end{bmatrix}}$$

  i List who each person can send direct messages to.

  ii Explain why the leading diagonal is all zeros.

  iii Draw a communication diagram showing the communication links given in the matrix.

  iv How could Zoe get a message to Chloe in two steps?

  i Look at each row in order.

    '1' means the person can send a direct message.

  ii Refer to redundant links.

  iii Draw a diagram with arrows that match the list of possible direct messages.

  iv Find how the message could be passed on in two steps.

b **i** Write the communication matrix for the following communication diagram.

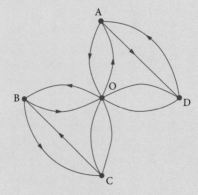

  ii Explain why the matrix is symmetric by referring to the communication diagram.

i Set up a 5 × 5 binary matrix where
'1' indicates communication and
'0' indicates non-communication.

ii Refer to the direction of the arrows
in the communication diagram.

MATCHED EXAMPLE 21	Working with two-step communication
Steps	**Working**

For the communication matrix representing the connections between four computers, find the following.

$$M = \text{from} \begin{array}{c} \\ P \\ Q \\ R \\ S \end{array} \overset{\displaystyle to}{\begin{bmatrix} P & Q & R & S \\ 0 & 1 & 1 & 0 \\ 1 & 0 & 0 & 1 \\ 1 & 0 & 0 & 1 \\ 0 & 1 & 1 & 0 \end{bmatrix}}$$

a The number of ways Q can connect with R by connecting directly to one other computer.

b The list of all the two-step connections from Q to R.

c The total number of redundant two-step connections.

d The list of redundant two step connections from P to P.

e The total number of one-step and two-step connections from S to R.

a Find M^2 using CAS and read the number of two-step connections from the matrix.	
b Use M to find the two-step connections.	
c Sum the values in the lead diagonal of M^2.	
d Use M^2 to find the number of redundant two-step connections. Use M to find the redundant two-step connections.	
e Find $M + M^2$, using CAS if necessary, and read the number from the matrix.	

For each of the following, find

i the dominance matrix

ii the total dominance scores

iii the ranking of the participants and overall winner.

Steps	Working

a Alex, Bethany, Chen, David and Ella competed in a small chess tournament in their town. The results are shown in the following dominance diagram, where an arrow indicates which player defeated the other.

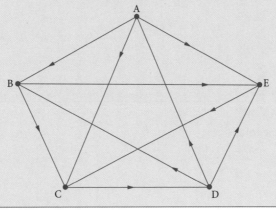

i Use the direction of the arrows in the diagram to construct a square binary matrix M, where dominance is shown by a 1.

ii Find $M + M^2$ using CAS and add each row to find the total dominance scores.

iii Rank from 1 to 5. Ties are given equal ranking, and the ranking of positions below the ties is not affected by the ties.

b Pearl, Quinn, Robin, Sarah and Tina played off against each other in a series of one-on-one badminton games. In these games:

- Pearl beat Robin and Sarah
- Quinn beat Pearl, Robin, Sarah and Tina
- Robin beat Sarah and Tina
- Sarah beat Pearl and Tina.

i Construct a square binary matrix M, where dominance is shown by a '1'.

ii Find $M + M^2$ using CAS and add each row to find the total dominance scores.

iii Rank from 1 to 5. Ties are given equal ranking, and the ranking of positions below the ties isn't affected by the ties.

MATCHED EXAMPLE 23 | Working with dominance matrices

SB

p. 461

The dominance matrix shows the result of each match between four teams, P, Q, R and S, in a round-robin tournament.

$$M = Winner \quad \begin{array}{c} \\ \\ P \\ Q \\ R \\ S \end{array} \begin{array}{c} \quad Loser \\ \begin{array}{cccc} P & Q & R & S \end{array} \\ \begin{bmatrix} c & 0 & 1 & 0 \\ 1 & 0 & 1 & b \\ 0 & a & 0 & 1 \\ 1 & 1 & 0 & 0 \end{bmatrix} \end{array}$$

Steps	Working

a Complete the dominance matrix by including the values for a, b and c.

The leading diagonals in dominance matrices are all zeros.

Since dominance is one way, if P to Q is '1', then Q to P is '0', and if P to Q is '0', then Q to P is '1'.

b The result of one game not involving Q in the tournament is in dispute. If the result of the game is reversed, team Q would be declared the clear overall winner. Which game is it?

   1 List the games that could be in dispute.

   2 For each of the reversed result options:
- state the reversed result
- find the dominance matrices
- calculate the total dominance scores
- state the overall winner.

   3 State the answer.

TRANSITION MATRICES

MATCHED EXAMPLE 1 | Constructing transition matrices from diagrams

Construct transition matrices for each of the following.

Steps	Working

a For this transition diagram showing changes from one city to the next

 i find p, q, r and s

 ii construct the matching transition matrix.

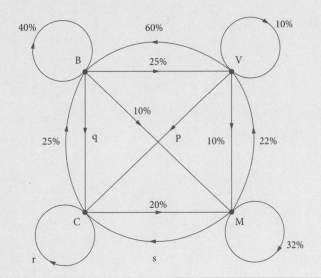

 i All the arrow percentages *from* a single point add up to 100%.

 Solve for the unknowns, using CAS if necessary.

 ii Convert all the percentages to decimals and construct the matrix.

This city

$\quad\quad B \quad\quad V \quad\quad M \quad\quad C$

$$\begin{bmatrix} & & & \\ & & & \\ & & & \\ & & & \end{bmatrix} \begin{matrix} B \\ V \\ M \\ C \end{matrix} \text{ Next city}$$

b Construct a transition matrix for the following situation. In a game of rummy, if a player wins the current game, there is a 25% chance he wins the next game. If he loses the current game, then there is a 60% chance he might lose the next.

1 Set up a 2×2 matrix using W for 'win' and L for 'lose' with a transition from 'This game' to 'Next game'. Enter the percentages from the question as decimals.

2 Enter the remaining elements of the matrix by using the fact that the columns of a transition matrix must add up to 1.

Add these to the matrix you have written for step 1.

MATCHED EXAMPLE 2 | Constructing transition diagrams from matrices

Steps	**Working**

Given the following incomplete transition matrix and matching transition diagram,

a find *a*, *b* and *c* to complete the transition matrix

b complete the matching transition diagram.

This month

$$\begin{array}{ccc} P & Q & R \end{array}$$

$$\begin{bmatrix} a & 0.70 & 0.48 \\ 0.63 & 0.30 & b \\ 0.20 & c & 0.52 \end{bmatrix} \begin{array}{l} P \\ Q \\ R \end{array} \text{ Next month}$$

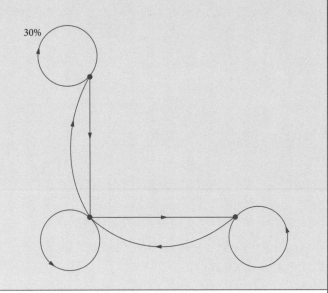

a 1 Each column of a transition matrix must add up to 1. Solve for the unknowns, using CAS if necessary.

2 Complete the transition matrix.

b 1 Use the transition matrix to label the points *P*, *Q* and *R* in the diagram.

Which transition is 0.3?

Which are the two points where there is no transition?

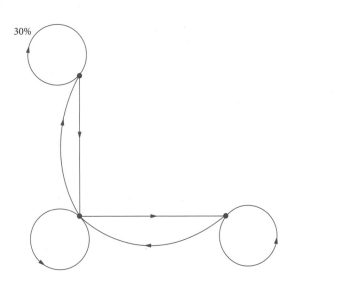

2 Complete the transition diagram from the transition matrix.   Add the remaining percentages to the diagram in step 1.

MATCHED EXAMPLE 3 | Interpreting transition matrices

A group of Australian pelicans migrates to three different places, P, Q and R, during the breeding season.

The Australian pelicans change their location each season according to the transition matrix T where

$$\begin{array}{c} \quad\quad\text{This season} \\ \quad P \quad\ Q \quad\ R \\ T = \begin{bmatrix} 0.53 & 0.26 & 0.35 \\ 0.17 & 0.42 & 0.15 \\ 0.30 & 0.32 & 0.50 \end{bmatrix} \begin{array}{l} P \\ Q \\ R \end{array} \text{Next season} \end{array}$$

This season there are 2500 Australian pelicans in location P, 1700 Australian pelicans in location Q and 3200 Australian pelicans in location R.

a How many Australian pelicans in location P this season are expected to be in location Q next season?

b How many Australian pelicans in location Q this season are expected to remain in location Q next season?

c How many Australian pelicans are expected be in location R next year?

d If 52% of the Australian pelicans were in location R one season, what percentage are expected to be in location P the next season?

Steps	Working
a Locate the relevant element in the transition matrix and multiply by the number of Australian pelicans.	
b Locate the relevant element in the transition matrix and multiply by the number of Australian pelicans.	
c Locate the relevant elements in the transition matrix, multiply by the number of Australian pelicans in each case and add.	
d Locate the relevant element in the transition matrix and multiply by the percentage.	

MATCHED EXAMPLE 4 Using permutation matrices for transitions

Alex has five shirts in five different colours, black (B), green (G), white (W), yellow (Y), and orange (O). Alex selected the black shirt to wear on the first day of the month. For the remaining days of the month, Alex selects his shirt according to the following permutation matrix, P.

$$P = \begin{array}{c} & \begin{array}{ccccc} B & G & W & Y & O \end{array} \\ \begin{array}{c} \\ \\ \\ \\ \\ \end{array} & \left[\begin{array}{ccccc} 0 & 0 & 0 & 1 & 0 \\ 0 & 0 & 1 & 0 & 0 \\ 1 & 0 & 0 & 0 & 0 \\ 0 & 1 & 0 & 0 & 0 \\ 0 & 0 & 0 & 0 & 1 \end{array}\right] \begin{array}{c} B \\ G \\ W \\ Y \\ O \end{array} \end{array}$$

This day above; Next day at right.

a Explain why P is a transition matrix.

b What does the '1' in column G and row Y mean?

c What colour shirt did Alex wear on the fourth day of the month?

Steps	Working
a Is it a square matrix? Does each column add up to 1?	
b A '1' indicates that the transition occurs 100% of the time.	
c The '1's indicate 100% certainty in choices. Use the transition matrix to list the selections starting from the first day.	

MATCHED EXAMPLE 5 | Using the state matrix recurrence relation

SB
p. 496

A group of people who have breakfast cereal switch between Loopy Loops and Crunchy Flakes every week according to the following matrix.

This week

$$T = \begin{array}{c c} & \begin{matrix} LL & CF \end{matrix} \\ \begin{bmatrix} 0.90 & 0.65 \\ 0.10 & 0.35 \end{bmatrix} & \begin{matrix} LL \\ CF \end{matrix} \end{array} \quad \text{Next week}$$

If we know that this week there are 450 people who buy Loopy Loops and 320 people who buy Crunchy Flakes

Steps	Working
a find the matrix recurrence relation that generates the sequence of state matrices	
Find the recurrence relation given by S_0 = initial state matrix, $S_{n+1} = TS_n$	
b use the recurrence relation to calculate the number of people predicted to buy Loopy Loops and Crunchy Flakes the next week	
1 Use the recurrence relation to find S_1.	
2 Write the answer, rounding to the nearest whole number if necessary.	
c show the calculations for the number of people predicted to buy Loopy Loops and Crunchy Flakes next week	
Write out the steps for the calculations from the matrix multiplication.	
d use the recurrence relation to calculate the number of people predicted to buy Loopy Loops and Crunchy Flakes the week after next.	
1 Use the recurrence relation and your unrounded answer for S_1 to find S_2.	
2 Write the answer, rounding to the nearest whole number if necessary.	

SB

p. 496

MATCHED EXAMPLE 6 | Interpreting the state matrix recurrence relation

A fish farm has three small ponds, P, Q and R, connected by small streams. The fish move between the three ponds each day according to the recurrence relation

$$S_0 = \begin{bmatrix} 3000 \\ 2200 \\ 4100 \end{bmatrix} \begin{matrix} P \\ Q \\ R \end{matrix}, S_{n+1} = TS_n$$

where

$$T = \begin{matrix} & \text{This day} \\ & \begin{matrix} P & Q & R \end{matrix} \\ \begin{bmatrix} 0.1 & 0 & 0 \\ 0.4 & 0.5 & 0 \\ 0.5 & 0.5 & 1 \end{bmatrix} & \begin{matrix} P \\ Q \\ R \end{matrix} \text{ Next day} \end{matrix}$$

and S_0 shows the numbers of fish at each pond this day.

Steps	Working
a Explain what the number 1 in the third row and third column of the transition matrix tells us.	
A '1' in the transition matrix indicates this transition is 100% certain to occur.	
b How many fish are expected to stay in the same pond the next day?	
1 Identify the proportions in the transition matrix that represent no change.	
2 Use the initial state matrix to calculate the total number of fish that do not move to a different pond.	
3 Write the answer, rounding to the nearest whole number if necessary.	
c What percentage of fish move to a different pond the next day? Round to the nearest percentage.	
1 Subtract the number of fish that do not move to a different pond the next day from the total number of fish.	
2 Calculate the percentage, rounding to the nearest percentage.	

9780170464055

MATCHED EXAMPLE 7 | Solving the state matrix recurrence relation

SB

p. 497

Find the regular transition matrix T by solving for the unknowns in the following state matrix recurrence relation.

$$S_0 = \begin{bmatrix} 120 \\ 220 \\ 300 \end{bmatrix}, S_{n+1} = TS_n$$

where

$$T = \begin{bmatrix} v & w & x \\ 0.3 & 0.2 & y \\ 0.5 & 0.3 & z \end{bmatrix} \quad S_1 = \begin{bmatrix} 254 \\ 230 \\ 156 \end{bmatrix}$$

Steps	Working
1 Use the fact that transition matrix columns add to 1.	
2 Write the recurrence relation for S_1.	
3 Multiply the matrix rows with unknowns and solve the equations.	
4 Write the transition matrix.	

SB

Using CAS 1:
Finding state
matrices using
the rule
p. 499

p. 500

MATCHED EXAMPLE 8 | Using the state matrix rule

A fleet of ships starts at one of the three seaports, Q, R or S. By the end of the month, the ships end up at one of the ports according to the transition matrix T.

This month

$$T = \begin{array}{c} \\ \\ \end{array} \begin{array}{ccc} Q & R & S \\ \end{array}$$

$$T = \begin{bmatrix} 0.2 & 0.5 & 0.3 \\ 0.2 & 0.4 & 0.25 \\ 0.6 & 0.1 & 0.45 \end{bmatrix} \begin{array}{l} Q \\ R \text{ Next month} \\ S \end{array}$$

At the start of a particular month, there are 1000 ships at port Q, 2200 ships at port R and 2650 ships at port S.

Steps	Working
a How many ships are at each port after three months?	
1 Use CAS and the state matrix rule $S_n = T^n S_0$.	
2 Write the answer.	
b How many ships are at each port after eight months?	
1 Use CAS and the state matrix rule $S_n = T^n S_0$.	
2 Write the answer.	

c After eight months, the movement of ships changes according to the following transition matrix. Calculate the number of ships expected to be at each port after 17 months.

This month

$$\begin{array}{c} \\ \end{array} \begin{array}{ccc} Q & R & S \\ \end{array}$$

$$Z = \begin{bmatrix} 0.4 & 0.3 & 0.5 \\ 0.2 & 0.5 & 0.25 \\ 0.4 & 0.2 & 0.25 \end{bmatrix} \begin{array}{l} Q \\ R \text{ Next month} \\ S \end{array}$$

1 Use CAS and the state matrix rule $S_n = T^n S_0$, making sure you always multiply the transition matrices on the left.

2 Write the answer.

MATCHED EXAMPLE 9 | Finding the equilibrium state matrix

A fleet of ships starts at one of the three seaports, Q, R or S. By the end of the month, the ships end up at one of the ports according to the transition matrix T.

This month

$$T = \begin{array}{c} \quad \begin{array}{ccc} Q & R & S \end{array} \\ \begin{bmatrix} 0.2 & 0.5 & 0.3 \\ 0.2 & 0.4 & 0.25 \\ 0.6 & 0.1 & 0.45 \end{bmatrix} \begin{array}{l} Q \\ R \\ S \end{array} \text{Next month} \end{array}$$

At the start of a particular month, there are 1000 ships at port Q, 2200 ships at port R and 2650 ships at port S.

Steps	Working
a Show that there will be an equilibrium state matrix.	
Does T or a power of T have any zero elements? Check T first, then T^2, if necessary.	
b Find the equilibrium state matrix.	
1 Use CAS and the rule $S^n = T^n S_0$ for two large consecutive values of n.	
2 Are the two state matrices the same?	
c How many ships will be at each port in the long term?	
Read from the equilibrium state matrix.	
d What percentage of ships are at port Q in the long term? Round your answer to the nearest percentage.	
Calculate from the equilibrium state matrix.	

MATCHED EXAMPLE 10 Finding long-term trends with transition matrices

Students at a school in the age group of 16–18 years either use the internet at home (H), work (W) or school (S) to do their homework assignments. The transition matrix shows how this changes from day to day.

$$
\begin{array}{c}
\quad\quad\quad \text{This day} \\
\quad\quad H \quad\ W \quad\ S \\
T = \begin{bmatrix} 0.5 & 0.3 & 0.1 \\ 0 & 0.5 & 0.4 \\ 0.5 & 0.2 & 0.5 \end{bmatrix} \begin{array}{l} H \\ W \\ S \end{array} \ \text{Next day}
\end{array}
$$

a Show that there will be an equilibrium state matrix.

b What percentage of students are expected to use the internet at home each day in the long term. Round your answer to the nearest percentage.

c If there are 400 students in the age group of 16–18 years at the school, find the equilibrium state matrix and hence find how many students are expected to use the internet at work in the long term.

d The school assumes the same transition matrix applies the following year. If in the long term 520 students are expected to use the internet at work, what would be the expected number of students that will use the internet at their school?

e The school realises the following year that a different transition matrix R applies. Describe the long-term trends shown by

$$
\begin{array}{c}
\quad\quad\quad \text{This day} \\
\quad\quad H \quad\ W \quad\ S \\
R = \begin{bmatrix} 0.3 & 0 & 0.6 \\ 0 & 1 & 0.2 \\ 0.7 & 0 & 0.2 \end{bmatrix} \begin{array}{l} H \\ W \\ S \end{array} \ \text{Next day}
\end{array}
$$

Steps	Working
a 1 Does T or a power of T have any zero elements? Check T first, then T^2, if necessary.	
b 1 Find T^n for two large consecutive values of n.	
2 Convert the required decimal to a rounded percentage.	

c **1** Create an initial state matrix from the total given.

2 Use CAS and the state matrix rule $S_n = T^n S_0$ for two large consecutive values of n.

3 Are the two state matrices the same?

4 Read from the equilibrium state matrix.

d **1** Use T^n for a large value of n and the given number of expected students using the internet from work to find the total number of students.

2 Use the total student number found to calculate the number of students expected to use the internet from school.

e **1** Find R^n for a large value of n.

2 Read from the matrix.

9780170464055

A scrabble club in a city assesses its members based on the number of games played by them in a year. After each year, members are classified as Expert (E), Master (M), Grandmaster (S) and members who leave (L) the club when they achieve the highest title. Members cannot be assessed at a level lower than their title. In the recurrence relation shown

- the initial state matrix S_0 shows the number of members at each level immediately before the first assessment of the year
- the transition matrix T contains the percentages of members who are expected to change their title after each assessment and members who are expected to leave
- the matrix J contains the number of members who will join the club after each assessment.

$$S_0 = \begin{bmatrix} 100 \\ 50 \\ 60 \\ 0 \end{bmatrix} \begin{matrix} E \\ M \\ G \\ L \end{matrix}, S_{n+1} = TS_n + J \qquad \text{where} \qquad T = \begin{bmatrix} 0.2 & 0 & 0 & 0 \\ 0.5 & 0.45 & 0 & 0 \\ 0.2 & 0.15 & 0.8 & 0 \\ 0.1 & 0.4 & 0.2 & 1 \end{bmatrix} \begin{matrix} E \\ M \\ G \\ L \end{matrix} \text{After assessment}$$

Before assessment
$E \quad M \quad G \quad L$

$$\text{and } J = \begin{bmatrix} 60 \\ 30 \\ 25 \\ 0 \end{bmatrix} \begin{matrix} E \\ M \\ G \\ L \end{matrix}$$

Steps	Working
a Find the number of members with an Expert title after the first assessment.	
1 Use the recurrence relation $S_0 =$ initial state matrix, $S_{n+1} = TS_n + J$ to find S_1 **2** Find the answer from S_1.	
b Find the number of students with a Master title after the second assessment.	
1 Use the recurrence relation to find S_2 **2** Find the answer from S_2.	

MATCHED EXAMPLE 12 | Restocking and culling with a constant total

A poultry farmer decides to raise quails for eggs, and he starts with 5000 eggs, 850 young quails, and 150 adult quails. The following could happen in a year:

- eggs (E) could die (D) or they could live and become young quails (Y)
- young quails (Y) could die (D) or they could live and become adult quails (A)
- adult quails (A) could die (D) or they could live for a while but will eventually die.

From one year to the next, this situation can be represented by the recurrence relation

$$S_0 = \begin{bmatrix} 5000 \\ 850 \\ 150 \\ 0 \end{bmatrix} \begin{matrix} E \\ Y \\ A \\ D \end{matrix} \quad S_{n+1} = TS_n + Q,$$

where

This year

$$T = \begin{bmatrix} E & Y & A & D \\ 0 & 0 & 0 & 0 \\ 0.2 & 0 & 0 & 0 \\ 0.2 & 0.25 & 0.5 & 0 \\ 0.6 & 0.75 & 0.5 & 1 \end{bmatrix} \begin{matrix} E \\ Y \\ A \\ D \end{matrix} \quad \text{Next year}$$

The farmer wants to buy or sell enough young and adult quails so that their numbers in the farm remain the same each year. The farmer also assumes that the female adult quails will lay 5000 eggs each year and that all the dead quails and eggs are removed. How many young and adult quails will the farmer buy or sell each year?

Steps	Working
1 The state matrices will remain the same each year.	
2 Substitute the known matrices into the recurrence relation.	
3 Solve for Q.	
4 Find the answer from Q.	

MATCHED EXAMPLE 13 | Solving for the restocking/culling matrix

A new restaurant analyses the choices its customers make between four main courses: crab (C), venison (V), swordfish (S) and lobster (L). S_n shows the number of customers who chose each course in the week n since the restaurant opened. S_3 and S_4 are given below. The matrix T shows how customers are expected to change their choice from week to week.

This week

$$S_3 = \begin{bmatrix} 720 \\ 520 \\ 190 \\ 130 \end{bmatrix} \begin{matrix} C \\ V \\ S \\ L \end{matrix} \qquad S_4 = \begin{bmatrix} 740 \\ 240 \\ 320 \\ 620 \end{bmatrix} \begin{matrix} C \\ V \\ S \\ L \end{matrix} \qquad T = \begin{bmatrix} 0.2 & 0.4 & 0.4 & 0.5 \\ 0.25 & 0.1 & 0.3 & 0.1 \\ 0.2 & 0.1 & 0.2 & 0.2 \\ 0.35 & 0.4 & 0.1 & 0.2 \end{bmatrix} \begin{matrix} C \\ V \\ S \\ L \end{matrix} \text{ Next week}$$

$$\begin{matrix} C & V & S & L \end{matrix}$$

The destinations chosen can be modelled by $S_{n+1} = TS_n + B$

a Find B.

b Explain the meaning of the negative element in B.

c Find S_5.

Steps	Working
a Write the recurrence relation in terms of the matrices given in the question. Solve for B.	
b A negative means removing something.	
c Use the restocking/culling recurrence relation.	

MATCHED EXAMPLE 14 | Using female birth and survival rate tables

A population of a particular species of squirrels consists of

- 40 female squirrels less than two years old of which 40% will survive to be two
- 30 two-year-old female squirrels of which 60% will survive to be four
- 20 four-year-old female squirrels of which none will survive to be six.

The females in the population have the following average yearly birth rates:

- Female squirrels less than two years old give birth on average to 5.4 squirrels.
- Female four-year-old squirrels give birth on average to 10.2 squirrels.
- Female six-year-old squirrels give birth on average to 6.2 squirrels.

Assume that half the population is female and half the births are female.

Steps	Working

a Explain why for the female birth and survival rate table for this information we need to halve the squirrels birth rates given.

Check which of the figures given apply to females only and which apply to both males and females.

b Draw up a female birth and survival rate table that includes the initial number of female squirrels.

Make sure that all the entries apply to females only. Convert percentage survival rates to decimals.

Age (years)	0–<2	2–<4	4–<6
Initial number			
Birth rate			
Survival rate			

c Write a calculation from the table that will show how many two-year-old female squirrels are expected to survive to be four.

Multiply the initial number and survival rate from the table. Round to the nearest whole number.

d Write a calculation from the table that will show how many female squirrels are expected to be born after two years.

Multiply the initial numbers of squirrels by each of their birth rates and add them together. Round to the nearest whole number.

e Write the Leslie matrix L that represents the information in the table.

Write the birth rates as the first row.
Write the survival rates without the last zero as the subdiagonal.

Write zeros everywhere else.

SB

p. 532

Data from a study of the four-year lifespan of a particular species of rodent population has been written as the following initial state matrix S_0 and Leslie matrix L.

$$S_0 = \begin{bmatrix} 350 \\ 430 \\ 500 \\ 250 \end{bmatrix} \begin{matrix} 0\text{–}<1 \\ 1\text{–}<2 \\ 2\text{–}<3 \\ 3\text{–}<4 \end{matrix} \qquad L = \begin{bmatrix} 0 & 12.5 & 12 & 10.5 \\ 0.2 & 0 & 0 & 0 \\ 0 & 0.3 & 0 & 0 \\ 0 & 0 & 0.2 & 0 \end{bmatrix}$$

Steps	Working

a How do we know from the Leslie matrix that the rodents do not breed until they are one year old?

Look at the first birth rate.

b Show a calculation using a recurrence relation to find the number of three-year-old rodents after one year.

  1 Find S_1 using the Leslie matrix recurrence relation.

  2 Find the number from S_1, multiply it by 2 and then round it to the nearest whole number if necessary.

c Show a calculation using a recurrence relation to find the total number of rodents after two years.

  1 Find S_2 using the Leslie matrix recurrence relation.

  2 Multiply each of the elements in S_2 by 2, rounding to the nearest whole number if necessary, and add them together.

  3 Write the answer.

d Find the number of three-year-old rodents after five years.

  1 Use the Leslie matrix rule.

  2 Find the number from S_5, multiply it by 2 and then round it to the nearest whole number if necessary.

MATCHED EXAMPLE 16 | Finding long-term Leslie matrix trends

For each of the following Leslie matrices L where $S_0 = \begin{bmatrix} 5000 \\ 0 \\ 0 \end{bmatrix}$, show whether in the long term the populations will increase, decrease or cycle.

a $L = \begin{bmatrix} 0 & 0 & 1000 \\ 0.15 & 0 & 0 \\ 0 & 0.02 & 0 \end{bmatrix}$ **b** $L = \begin{bmatrix} 0 & 0 & 334 \\ 0.15 & 0 & 0 \\ 0 & 0.02 & 0 \end{bmatrix}$ **c** $L = \begin{bmatrix} 0 & 0 & 84 \\ 0.15 & 0 & 0 \\ 0 & 0.02 & 0 \end{bmatrix}$

Steps	Working
a **1** Since L is 3×3, find S_3 using the Leslie matrix rule $S_n = L^n S_0$.	
2 Compare S_3 and S_0.	
b **1** Since L is 3×3, find S_3 using the Leslie matrix rule $S_n = L^n S_0$.	
2 Compare S_3 and S_0.	
c **1** Since L is 3×3, find S_3 using the Leslie matrix rule $S_n = L^n S_0$.	
2 Compare S_3 and S_0.	

MATCHED EXAMPLE 17 | Finding long-term Leslie matrix growth rates

For the following initial state matrix S_0 and Leslie matrix L, find

a the long-term growth rate per time period to two decimal places

b the percentage increase or decrease per time period to the nearest percentage.

$$S_0 = \begin{bmatrix} 7200 \\ 6400 \\ 0 \end{bmatrix} \quad L = \begin{bmatrix} 0 & 3 & 2 \\ 0.25 & 0 & 0 \\ 0 & 0.3 & 0 \end{bmatrix}$$

Steps	Working
a 1 Use the rule $S_n = L^n S_0$ for two large consecutive values of n. Round to one decimal place.	
2 Find the ratios: $\dfrac{\text{Each of the elements in } S_{n+1}}{\text{Matching element in } S_n}$	
3 Write the long-term growth rate per time period g to two decimal places.	
b 1 Calculate $(1 - g) \times 100\%$.	
2 Is $g > 1$ (increase) or $g < 1$ (decrease)? Write the answer to the nearest percentage.	

UNDIRECTED GRAPHS

SB

p. 557

MATCHED EXAMPLE 1 | Identifying isomorphic graphs

For each of the following pairs of graphs, state whether or not they are isomorphic and give a reason for your answer.

a

b

c

Steps	Working

Do the graphs have the same number of vertices?

Do the graphs have the same number of edges?

Do the graphs show *exactly* the same connections?

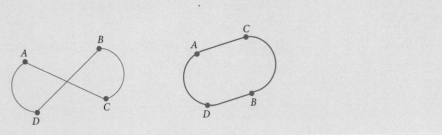

For each of the following graphs

   i count and list the vertices, edges and faces

   ii show that the degree sum is twice the number of edges.

a

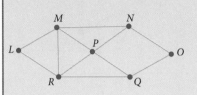

b

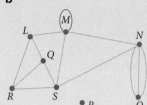

Steps	Working
a **i** Count and list the number of vertices and edges. Count and list the number of enclosed regions plus the region outside the graph. **ii** Find the degree of each vertex and sum them. Show that multiplying the number of edges by 2 gives the same result.	

Vertex	L	M	N	O	P	Q	R	Sum
Degree								

Steps	Working
b **i** **1** Redraw the graph to uncross the intersecting edges that have no vertex at the point of intersection. **2** Count and list the number of vertices and edges. Count and list the number of enclosed regions, plus the region outside the graph. **ii** Find the degree of each vertex and sum them. Show that multiplying the number of edges by 2 gives the same result.	

Vertex	L	M	N	O	P	Q	R	S	Sum
Degree									

MATCHED EXAMPLE 3 · Representing road systems as graphs

For the road system showing how roads connect the three towns
Fista (*F*), Sayge (*S*) and Chinko (*C*), find all the routes between the
towns that do not go through one of the other towns, and hence, draw
a graph representing the road connections.

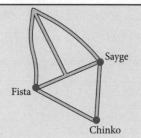

Steps	Working
1 Find the loops by identifying a route from a town back to the same town that does not go through another town.	
2 Find the edges by identifying routes between two towns that do not go through another town.	

3 Draw the graph.

MATCHED EXAMPLE 4 Representing maps as graphs

The city of Lakeville is divided into five suburbs labelled as
A to *E* on the map. A lake in the middle of the city is shown
on the map. Draw a graph showing the land connections
between the five suburbs.

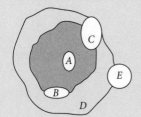

Steps	Working
1 List all of the land connections.	
Let *A*, *B*, *C*, *D* and *E* be the vertices of the graph. The land connections are the edges. Draw the vertices and connect them with the number of edges.	

MATCHED EXAMPLE 5 | Identifying bridges

How many bridges are there in each of these connected graphs? Copy the graphs and indicate the bridges by drawing them in blue.

a
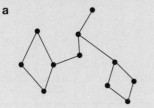

b
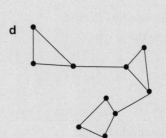

c
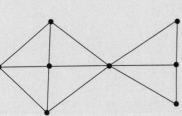

d

Steps	Working
Decide which edges will make the graph disconnected if you delete them.	

MATCHED EXAMPLE 6 | Verifying Euler's formula

For the graph shown

a redraw it to show it is a planar graph

b state whether or not it is a connected graph, giving a reason

c verify that Euler's formula works or show that it doesn't.

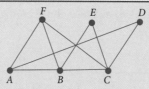

Steps	Working
a To uncross the edges, move edges *FC* and *AD* around the outside of the graph.	
b Use the definition of a connected graph.	
c Count the number of vertices, faces and edges, and substitute into Euler's formula to see if the result is 2.	

MATCHED EXAMPLE 7	Using Euler's formula

| **a** A connected planar graph has 11 edges and 8 faces. How many vertices does it have? |
| **b** A connected planar graph has 12 vertices and 9 faces. How many edges does this graph have? |

Steps	Working
a Substitute the known values into Euler's formula and solve to find the number of vertices, v.	
b Substitute the known values into Euler's formula and solve to find the number of edges, e.	

MATCHED EXAMPLE 8 | Identifying simple graphs, complete graphs and subgraphs

For each of the following graphs, state, giving reasons for your answers, whether it is a

    i simple graph

    ii complete graph

    iii subgraph of

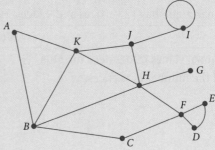

a

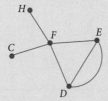

b

c

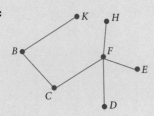

d

Steps	Working
i Does the graph have any loops or multiple edges?	
ii Is it a simple graph where every vertex is connected by one edge to every other vertex?	
iii Does the graph contain *only* vertices and edges from the original graph?	

MATCHED EXAMPLE 9 Classifying walks shown on a graph

For each of the following walks, state whether it is a trail, path, circuit, cycle or walk only, and give a reason for your answer.

a
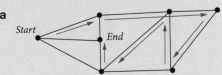

b
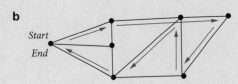

c
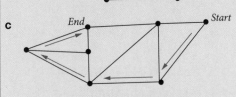

d

e
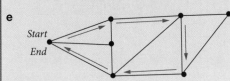

Steps	Working

Use the Walk classification chart to ask three questions, in this order, for each one:

1 Does the walk have repeated edges?

2 Does the walk have repeated vertices?

3 Does the walk start and finish at the same vertex?

MATCHED EXAMPLE 10 | Classifying walks from a list of vertices

Joe is trekking along Jebel Jais. For each of the following walks, state whether it is a trail, path, circuit, cycle or walk only, and give a reason for your answer.

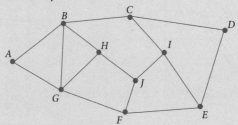

a A-B-H-J-F-G-A

b A-B-C-D-E

c A-G-H-J-F-G-H-B-A

d A-B-G-H-B-C

e F-J-H-G-B-H-J-F

f A-B-G-H-J-F-G-A

Steps	Working
Use the Walk classification chart to ask three questions, in this order, for each one: **1** Does the walk have repeated edges? **2** Does the walk have repeated vertices? **3** Does the walk start and finish at the same vertex?	

MATCHED EXAMPLE 11 | Classifying Eulerian and Hamiltonian walks

a For each of these graphs, give a reason why an Eulerian trail, an Eulerian circuit or neither exists, and describe three walks for each one that exists.

i

ii

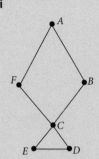

iii

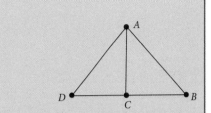

b For each of the graphs in part **a**, describe a walk that is a Hamiltonian path and a walk that is a Hamiltonian cycle for each one that exists.

Steps	Working
a i Count how many vertices have an odd degree.	
ii 1 Count how many vertices have an odd degree.	
2 An Eulerian circuit includes each edge once only and starts and finishes at the same vertex.	
iii 1 Count how many vertices have an odd degree.	
2 An Eulerian trail includes each edge once only and starts and finishes at vertices of odd degree.	
b A Hamiltonian path includes every vertex in a graph once only. A Hamiltonian cycle also starts and finishes at the same vertex. Trial and error is the only method that can be used.	

SB

p. 588

MATCHED EXAMPLE 12 | Representing graphs using adjacency matrices

Represent the graph using an adjacency matrix.

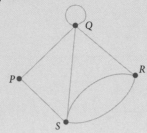
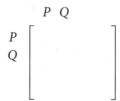

Steps	Working
1 Label the rows and columns of the matrix to match the graph.	$\begin{array}{cc} & P \quad Q \\ \begin{array}{c} P \\ Q \end{array} & \left[\begin{array}{cc} & \\ & \end{array}\right] \end{array}$
2 List the connections in terms of the number of edges between vertices.	
3 Fill in the matrix based on the number of edges. An edge from A to B is also an edge from B to A, so the matrix is symmetrical around the leading diagonal.	Write the elements in the matrix in Step 1.
4 Complete the matrix by writing 0 for all the remaining elements.	Write the elements in the matrix in Step 1.

MATCHED EXAMPLE 13 | Using adjacency matrices

For the following adjacency matrix representing
a planar graph:

$$
\begin{array}{c c} & \begin{array}{c c c c} A & B & C & D \end{array} \\ \begin{array}{c} A \\ B \\ C \\ D \end{array} & \left[\begin{array}{c c c c} 0 & 2 & 1 & 2 \\ 2 & 0 & 1 & 1 \\ 1 & 1 & 1 & 1 \\ 2 & 1 & 1 & 0 \end{array}\right] \end{array}
$$

Steps	Working
a How do we know it represents a connected graph?	
Check that every vertex is connected to at least one other vertex.	
b Find the degree of each vertex and, hence, find the degree sum of the graph.	
1 Sum the rows to find the degree of each vertex. Add an extra degree when there's a loop at the vertex.	
2 Sum the degrees of each vertex.	
c Does the graph have an Eulerian trail? Give a reason for your answer.	
A graph has an Eulerian trail if it has exactly 2 vertices of odd degree.	
d How many faces does the graph have?	
1 Find the number of vertices and edges. Number of vertices = number of rows Number of edges = sum of elements on and below the leading diagonal	
2 Since this is a connected planar graph, we can use Euler's formula to find the number of faces.	

e Draw the graph without any crossed edges.

Place the four vertices.

Use the elements on and below the leading diagonal to draw the edges of the graph.

MATCHED EXAMPLE 14 | Finding the shortest path by inspection

SB

p. 594

The network shows the travel times, in minutes, along a series of roads. Find the shortest time, in minutes, that it takes Peter to travel from his house to Clara's house by listing all the options.

Steps	Working
1 Add labels to the vertices.	
2 List the path options and calculate the total time of each option.	
3 Write the answer	

MATCHED EXAMPLE 15 | Finding the shortest path using Dijkstra's algorithm

p. 595

The network shows the travel times, in minutes, along a series of roads.

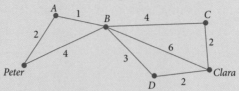

Steps	Working

a Find the shortest time, in minutes, that it takes Peter to travel from his house to Clara's house using Dijkstra's algorithm.

1 Box the starting vertex and add a value of 0.

2 For all the vertices connected to the starting vertex, add the value of each edge.

3 Box the smallest of the unboxed vertices.

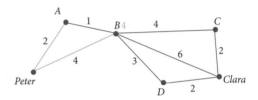

4 For all the unboxed vertices connected to the newly boxed vertex, add the value of each edge.

    If an unboxed vertex already has a value, replace the existing value with the new value if the new value is smaller.

5 Box the smallest of the unboxed vertices.

    Repeat the steps for the newly boxed value until the end vertex is boxed.

6 Write the next steps in the above graph.

Write the next step in the above graph.

7

8

9 The value in the end box is the shortest time.

b Use backtracking to find the shortest time path and draw it on the map.

Start from the end vertex and choose the boxed value that gives the correct value when you subtract the edge.

$8 - 2 \neq 7$, so don't move to C.

$8 - 2 = 6$, so move to D.

Continue this until you get back to the start.

Draw the path you've taken.

MATCHED EXAMPLE 16 | Identifying spanning trees

Which of the following graphs are a spanning tree of the graph shown? For those that are spanning trees, verify that the number of edges is one less than the number of vertices. For those that aren't spanning trees, give a reason.

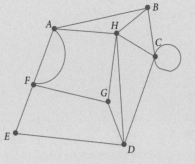

a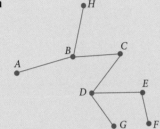

b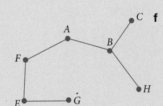

c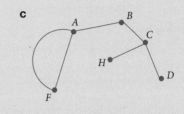

d

e

f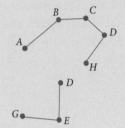

Steps	Working
Is it connected?	
Does it have all the vertices of the original graph?	
Does it have no loops?	
Does it have no multiple edges?	
Does it have no cycles?	

p. 604

MATCHED EXAMPLE 17 | Finding minimum spanning trees by inspection

Find all the spanning trees for the network shown, and, hence, find the total weight of the minimum spanning tree.

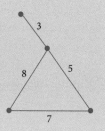

Steps	Working
1 Work out how many edges need to be removed from the graph to create a spanning tree.	

- The number of edges in a spanning tree is always one less than the number of vertices.

- A spanning tree has the same number of vertices as the original graph.

2 Remove each edge in turn to see which options result in a spanning tree. Calculate the total weight of each spanning tree and find the one with the smallest total weight.

Use Prim's algorithm to find the minimum spanning tree for the weighted graph shown, and, hence, find the total weight of the minimum spanning tree.

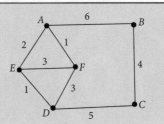

SB

p. 605

9

Steps

1 Start at any vertex and choose the edge with the lowest weight connected to this vertex.

2 Look at all the edges connecting to the vertices you've chosen so far (*not just the last vertex connected*) and choose the edge with the lowest weight that doesn't connect to a vertex already in the tree. If there are edges with equal lowest weights, choose one of them.

3 Repeat step **2** until all the vertices in the graph are included in the tree.

Working

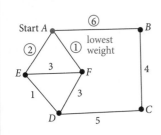

CHAPTER

10 DIRECTED GRAPHS

p. 628

MATCHED EXAMPLE 1	Drawing a directed graph from an activity table

Draw the directed graph for the activity table.

Activity	Activity time (hours)	Immediate predecessor
A	6	–
B	4	A
C	2	A
D	5	A
E	3	B
F	1	E, D
G	5	F, C

Steps	Working

1 Activity A has no predecessors.

Label the first vertex 'start' and from this a directed edge labelled A, 6.

2 Activities B, C and D are all preceded by A. From the vertex at the end of edge A, draw three directed edges labelled B, 4, C, 2 and D, 5.

3 Activity E is preceded by B. From the vertex at the end of edge B, draw an edge labelled E, 3.

Activity F is preceded by activities E and D. Draw an edge and connect the edges E and D to it. From this vertex, draw an edge labelled F, 1.

4 Activity G is preceded by activities F and C. Draw an edge and connect the edges F and C to it. From this vertex, draw an edge labelled G, 5.

5 The final vertex is drawn at the end of edge G and labelled 'finish'.

9780170464055

MATCHED EXAMPLE 2 | Finding the reachability of a vertex

In the directed graph, determine which vertices are **not reachable** from vertex A.

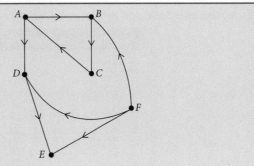

Steps	Working
1 Trace the paths from vertex A to every other vertex.	
The **vertices** that can be reached are coloured **blue** and the paths have **blue edges**.	
2 If a path does not exist from vertex A to a vertex, it is **not** reachable.	

MATCHED EXAMPLE 3 | Drawing a directed graph with dummy activities

Draw the directed graph for the project shown below, including dummy activities where required.

Activity	Activity time (hours)	Immediate predecessor
A	3	-
B	4	A
C	2	B
D	1	B
E	5	A, D, C

Steps	Working

1 Activity *A* has no predecessors.

Label the first vertex 'start' and from this draw directed edge labelled *A*, 3.

2 Activity *B* is preceded by activity *A*.

Draw a directed edge labelled *B*, 4 from the vertex at the end of activity *A*.

3 Activities *C* and *D* are preceded by activity *B*.

Draw two directed edges labelled *C*, 2 and *D*, 1 from the vertex at the end of activity *B*.

4 Activity *E* is preceded by activities *A*, *C* and *D*; however, two vertices cannot be connected by multiple edges. Two dummy activities must be included after activity *A* and activity *C*.

Draw a directed edge labelled *E*, 5 from the vertex at the end of activity *D*.

5 Draw a vertex labelled 'finish' at the end of activity *E*.

| MATCHED EXAMPLE 4 | Finding earliest start times |

Determine the earliest start time (EST) for each activity in the network shown. Activity times shown are in hours.

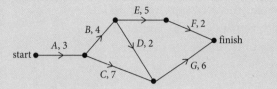

Steps	**Working**
1 Draw boxes above each vertex and enter 0 as the EST of the 'start' vertex. Double boxes are required for both activities B and C. Double boxes are required for both activities D and E.	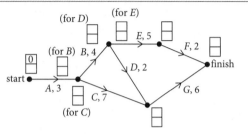
2 To find the EST for the second vertex, add the activity time of 3 to the previous EST. This gives an earliest start time for activities B and C as 3 hours.	Do this in the diagram in Step 1.
3 The EST for activity D and E are $3 + 4 = 7$. This is written in the top box at the start of activities D and E. This process is repeated: EST for activity $F = 7 + 5 = 12$	Do this in the diagram in Step 1.

4 To calculate the EST for the activity G, find the total for the paths containing activity C and the paths containing activity D. The highest figure obtained is the final EST.
 	
    Activity C: $3 + 7 = 10$
 	
    Activity D: $7 + 2 = 9$
 	
Therefore, 10 hours is the EST for activity G.
	Do this in the diagram in Step 1.

5 To calculate the EST for the finish, find the total for the paths containing activity F and the path containing activity G. The highest figure obtained is the final EST.
 	
    Activity F: $12 + 2 = 14$
 	
    Activity G: $10 + 6 = 16$
 	
Therefore, 16 hours is the final EST.
	Do this in the diagram in Step 1.

MATCHED EXAMPLE 5 Finding latest start times

Determine the latest start time (LST) for each activity in the network. Activity times shown are in hours.

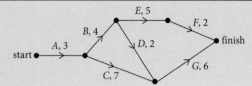

Steps	Working
1 First, calculate the EST for each activity. These were completed in Matched example 4.	
2 Work backwards from right to left. The LST for the 'finish' is equal to the EST.	
3 The LST for activity *F* is 16 − 2 = 14. This is written in the bottom box at the start of activity *F*. The LST for activity *G* = 16 − 6 = 10	
4 Calculate the LST for activities *D* and *E*. Activity *D*: 10 − 2 = 8 Activity *E*: 14 − 5 = 9	
5 Calculate the LST for activities *B* and *C*. If there are two LST values at the end of an activity, then the smallest value is used to calculate the LST. Therefore, use 8 hours to calculate the LST for the activity *B*. Activity *B*: 8 − 4 = 4 Activity *C*: 10 − 7 = 3 If there are two LST values at the end of an activity, then the smallest value is used to calculate the LST. Therefore, use 3 hours to calculate the LST for the start. The start LST = 3 − 3 = 0.	

Determine the earliest start time (EST) and latest start time (LST) for each activity in the network and hence determine the critical path. Activity times shown are in hours.

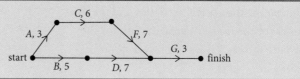

p. 638

Steps	Working
1 First calculate the EST for each activity.	
2 Calculate the LST for each activity.	
3 Activities where EST = LST are critical activities and are on the critical path.	

MATCHED EXAMPLE 7 | Finding float times for a directed graph

Determine the critical activities and the float times for the non-critical activities for the project shown. Times shown are in days.

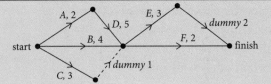

Steps	Working
1 First, calculate the EST and the LST for each activity.	
2 Identify the critical path. On the critical path, EST = LST.	
3 Calculate float times for the non-critical activities *B*, *C* and *F* using the formula $\qquad$ float = LST − EST	

MATCHED EXAMPLE 8 Applying crashing to a directed graph

a Determine the critical path and the minimum project completion time for the project shown. Activity times shown are in days.

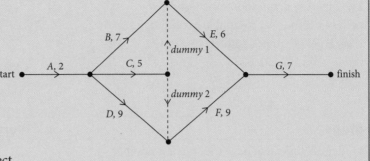

b Extra resources are used to speed up the original project resulting in activities *D*, *F* and *G* each being reducing by four days. Find the new critical path and the minimum project completion time.

Steps	Working
a 1 First, calculate the EST and the LST for each activity.	
2 On the critical paths, EST = LST. The minimum project completion time is EST or LST for the finish vertex.	
b 1 Reduce the times for activities *D*, *F* and *G* by four days.	
2 Calculate the EST and the LST for each activity with the new network.	
3 Identify the activities where EST = LST to find the critical path.	

MATCHED EXAMPLE 9 Finding the allocation from a bipartite graph

The bipartite graph shows the tasks that each of the four people is able to undertake. If each person must complete one task, find a valid allocation.

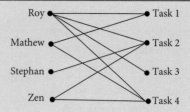

Steps	Working
1 Identify tasks that have the smallest number of links.	
2 Roy is the only person who can do task 3. Allocate Roy to task 3, then eliminate links from Roy to tasks 1, 2 and 4 and eliminate link from Zen to task 2.	
3 Task 1 can only be done by Mathew, so allocate Mathew to task 1, then eliminate links from Mathew to task 4.	
4 Allocate Zen to task 4, this was determined at step 1.	

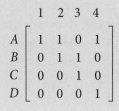
MATCHED EXAMPLE 10 | Drawing a bipartite graph from an adjacency matrix

Four workers A, B, C and D need to be allocated one task from task 1, task 2, task 3 and task 4. The workers are only qualified to perform certain tasks, and this is indicated in the following matrix. Draw a bipartite graph from the adjacency matrix and determine a valid allocation.

$$
\begin{array}{c}
\\ A \\ B \\ C \\ D
\end{array}
\begin{array}{c}
\begin{array}{cccc} 1 & 2 & 3 & 4 \end{array} \\
\left[\begin{array}{cccc}
1 & 1 & 0 & 1 \\
0 & 1 & 1 & 0 \\
0 & 0 & 1 & 0 \\
0 & 0 & 0 & 1
\end{array} \right]
\end{array}
$$

Steps	Working
1 Draw a connection for every 1 in the matrix. $A1$, $A2$, $A4$, $B2$, $B3$, $C3$ and $D4$	
2 Start with the tasks with the smallest number of links. Task 1 can only be done by A, then eliminate $A2$ and $A4$.	
3 Task 2 can only be done by B so allocate B to task 2 and remove any other link from B. Task 4 can only be done by D. Task 3 can only be done by C.	

MATCHED EXAMPLE 11 Applying stage one of the Hungarian algorithm

Three workers X, Y, and Z need to be allocated one task from task 1, task 2 and task 3. The time in hours that each worker takes to complete the tasks is shown in the table below.

	1	2	3
X	15	17	19
Y	10	11	14
Z	9	8	13

If the tasks must be completed in the minimum time, find how the tasks are assigned.

Steps	Working

1 Identify the smallest number in each row and subtract it from the other numbers in the same row.

    Row X subtract 15

    Row Y subtract 10

    Row Z subtract 8

	1	2	3
X			
Y			
Z			

2 For every column that does not have a zero value, choose the smallest number in the column and subtract it from every element in the same column.

    Column 1 has a zero

    Column 2 has a zero

    Column 3 subtract 4

	1	2	3
X			
Y			
Z			

3 Cover all the zeros with the smallest number of lines (horizontal or vertical).
The number of lines must equal the number of rows.

Do this in the table above.

4 The zeros indicate the allocations.
Draw a bipartite graph connecting
$X1$, $X3$, $Y1$, $Y3$ and $Z2$

A company has four employees who were given four different tasks.
The hours required by the employees to complete the tasks are
shown in the matrix, where the rows represent the employees
A, B, C and D and the columns represent the tasks 1, 2, 3 and 4.

$$\begin{array}{c} & \begin{array}{cccc} 1 & 2 & 3 & 4 \end{array} \\ \begin{array}{c} A \\ B \\ C \\ D \end{array} & \left[\begin{array}{cccc} 10 & 6 & 8 & 9 \\ 15 & 11 & 14 & 13 \\ 20 & 18 & 22 & 23 \\ 16 & 19 & 14 & 15 \end{array} \right] \end{array}$$

a Determine the best allocation of employees to minimise
the total time taken to complete the tasks.

b Find the total time taken to complete the four tasks.

Steps	Working

a 1 Perform a row reduction

Row A   subtract 6

Row B   subtract 11

Row C   subtract 18

Row D   subtract 14

	1	2	3	4
A				
B				
C				
D				

2 Perform a column reduction

Column 1   subtract 2

Column 4   subtract 1

	1	2	3	4
A				
B				
C				
D				

3 Cover the zeros

There are three lines and four rows.

As the number of lines does not equal
the number of lines does not equal
the number of rows, stage 2 of the Hungarian
algorithm must be used.

Do this in the table above.

4 The smallest uncovered number is 1. Add this
to every covered number but add 2 to the
numbers covered twice.

	1	2	3	4
A				
B				
C				
D				

5 The smallest value is 1, so subtract this from
every element in the matrix.

Now cover the zeros. There are now four lines
and four rows, so the allocation is complete.

	1	2	3	4
A				
B				
C				
D				

6 Draw the bipartite graph.

7 Determine the allocation.

One task has a single employee connected.

Employee D for task 3

This leaves

Employee B for task 4

It leaves

Employee A for task 2

And finally, this leaves

Employee C for task 1

b Look at the original matrix to determine the total time:

$A2 = 6$, $B4 = 13$, $C1 = 20$, $D3 = 14$

9780170464055

p. 663

The directed graph shows the flow capacities
in litres per minute. Determine

a the inflow capacity of vertex A.

b the outflow capacity of vertex A.

c the maximum flow out of vertex A.

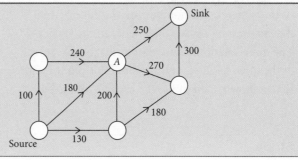

Steps

a The total inflow capacity of vertex A can be
found by adding the values on the edges
entering vertex A.

b The total outflow capacity of vertex A can
be found by adding the values on the edges
leaving vertex A.

c The maximum flow out of vertex A is the smaller
of inflow capacity and outflow capacity.

SB

p. 664

MATCHED EXAMPLE 14 | Finding the capacity of a cut

In the network, the numbers on the edges show
the maximum possible flow between the vertices.
The direction of the arrow indicates the direction
of the flow. A cut separating the sink from the source
is also shown. Determine the capacity of the cut.

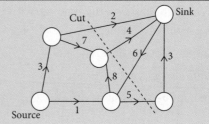

Steps	Working

Add the flows where the arrow crosses the cut in
the direction from the source side of the cut to
the sink side of the cut.

Flow 6 is not included as this flows in the
direction sink to source across the cut.

Finding the maximum flow of a network using cut capacities

Find the maximum flow for the network.

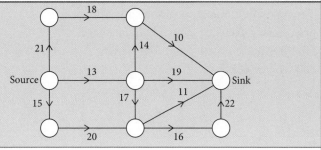

SB

p. 665

Steps	Working
1 Identify cuts that stop the flow from source to sink.	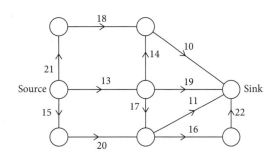
2 Find the capacity of each cut.	
3 Maximum flow = minimum cut	

MATCHED EXAMPLE 16 Applying Dijkstra's algorithm to directed graphs

The times, in minutes, to various tracks in an orienteering event are shown in the network. Find the quickest path from A to Z using Dijkstra's algorithm.

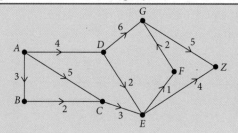

Steps	Working
1 Draw a box around the starting vertex A and enter a value of zero.	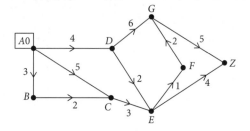
2 The boxed vertex A is connected to vertices B, C and D. Add the times on the connecting edges to the boxed value of 0 and write these totals next to the vertices in a different colour. Box the smallest value.	Do this in the diagram in step 1.
3 B is connected to C only: $3 + 2 = 5$, which is the same as the current vertex value, so we leave 5 as the value for C. Box the smallest value.	Do this in the diagram in step 1.
4 Consider all the unboxed vertices connected to the vertex D. Box the smallest value.	Do this in the diagram in step 1.
5 C is connected to E only: $5 + 3 = 8$, which is larger than the current vertex value, so we leave 6 as the value for E. Box the smallest value.	Do this in the diagram in step 1.

6 Consider all the unboxed vertices connected to the vertex E.

Box the smallest value.

Do this in the diagram in step 1.

7 F is connected to G only: $7 + 2 = 9$, which is smaller than the current vertex value, so we change 9 as the value for G.

Box the smallest value.

Do this in the diagram in step 1.

8 G is connected to Z only: $9 + 5 = 14$, which is larger than the current vertex value, so we leave 10 as the value for Z.

Box the smallest value.

Do this in the diagram in step 1.

9 Backtrack from vertex Z to vertex A.

Answers

Worked solutions available on Nelson MindTap.

CHAPTER 1

MATCHED EXAMPLE 1

a **i** Categorical   **ii** Ordinal

b **i** Numerical   **ii** Discrete

c **i** Categorical   **ii** Ordinal

d **i** Numerical   **ii** Continuous

e **i** Categorical   **ii** Nominal

MATCHED EXAMPLE 2

a

Juice	Frequency	Percentage
Apple	5	16.7%
Strawberry	9	30%
Orange	4	13.3%
Watermelon	6	20%
Carrot	3	10%
Pineapple	3	10%
Total	**30**	**100%**

b

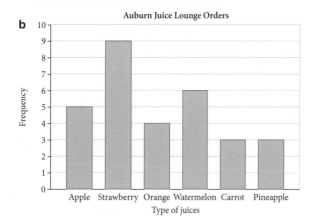

c

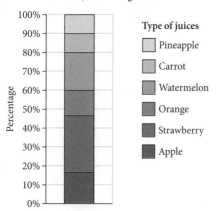

d The most frequently occurring category is 'Strawberry'.

MATCHED EXAMPLE 3

a **i** Modes = 3 and 9   **ii** Mean = 5.375

    iii Median = 5.5   **iv** Range = 8

b **i** Mode = 2.7°C

    ii Mean = 0.83°C

    iii Median = 1.8°C

    iv Range = 8.3°C

MATCHED EXAMPLE 4

a 45 people's heights range between 120 and 140 cm.

b The histogram is negatively skewed with no possible outliers.

c The modal interval is 150–<160 cm.

d The median is in the interval 140–<150 cm.

MATCHED EXAMPLE 5

a min = 4, Q_1 = 13.5, Q_2 = 21, Q_3 = 31, max = 56

b

MATCHED EXAMPLE 6

45 could be an outlier.

MATCHED EXAMPLE 7

a min = 19, Q_1 = 20, median = 23, Q3 = 26, max = 28

b 25% of employees scored more than 26.

c 50% of employees scored less than 23.

d 75% of employees scored between 19 and 26.

e 15

f Scores less than 11 would be considered outliers.

g No scores at the upper end would be considered outliers.

MATCHED EXAMPLE 8

a **i** 7.11   **ii** 7.1

b **i** 4.43   **ii** 4.4

c **i** 0.78   **ii** 0.78

d **i** 28 847.00   **ii** 29 000

e **i** 86.66   **ii** 87

f **i** 42 478.08   **ii** 42 000

g **i** 57.79   **ii** 58

MATCHED EXAMPLE 9

a **i** 2

    ii 7

    iii 60

b 19.5%

MATCHED EXAMPLE 10

a **i** mode = 21

  ii range = 5

  iii median = 21

  iv $Q_1 = 20$

  v $Q_3 = 22.5$

  vi IQR = 2.5

b The distribution is approximately skewed.

MATCHED EXAMPLE 11

a **i** Mode = 43

  ii Range = 46

  iii Median = 29

  iv $Q_1 = 19$

  v $Q_3 = 42$

  vi IQR = 23

b Since 2 isn't less than −15.5, it is not an outlier.

MATCHED EXAMPLE 12

a **i** mean = 3.1 toys

  ii median = 3 toys

  iii The mean and median are both appropriate measures of centre because the distribution is approximately symmetric with no outliers.

  iv standard deviation = 1.1 items

  v IQR = 2 items

  vi The standard deviation and IQR are both appropriate measures of spread because the distribution is approximately symmetric with no outliers.

b **i** mean = 26.3 cars

  ii median = 22.5 cars

  iii The median is an appropriate measure of centre because the distribution is skewed with no outliers.

  iv standard deviation = 12.0 cars

  v IQR = $Q_3 - Q_1 = 34 - 16.5 = 17.5$ cars

  vi The IQR is an appropriate measure of spread because the distribution is skewed with no outliers.

MATCHED EXAMPLE 13

a

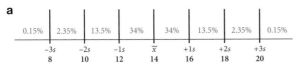

b 95% of lions have lifespan between 10 and 18 years.

c 16% of Lions have lifespan greater than 16 years.

d 84% of Lions have lifespan greater than 12 years.

e **i** 4.5 Lions would have lifespan lesser than 8 years.

  ii The zoo would expect 2991 lions to have lifespan between 8 and 20 years.

MATCHED EXAMPLE 14

a $z = 1$

b

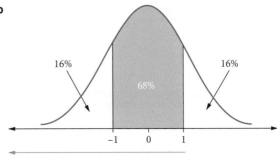

68% + 16% = 84%.

c The actual sale that day would be $600.

MATCHED EXAMPLE 15

a Computer Science

b Computer Science

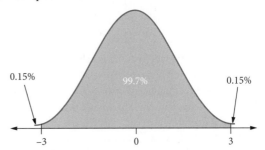

CHAPTER 2

MATCHED EXAMPLE 1

a **i** *Temperature of water* and *Number of fish*

  ii Both variables are numerical.

  iii The aim of the studies is to check if the temperature of water in the lake explains the number of fish in the lake.

So, *Temperature of water* is the explanatory variable.

b **i** *Hair serums* and *Hair growth*

  ii *Hair growth* is numerical and *Hair serums* is categorical.

  iii The hair serums are likely to affect the hair growth of a person.

So, *Hair serums* is the explanatory variable.

c **i** *Origin of tea leaves* and *Stimulating effect*

  ii Both variables are categorical.

  iii The origin of the tea leaves is likely to affect the stimulating effect. However, the stimulating effect is not likely to affect the origin of the tea leaves.

So, *Origin of tea leaves* is the explanatory variable.

d **i** *Workout intensity* and *Weight loss*

  ii *Workout intensity* is categorical and *Weight loss is numerical.*

  iii The experiment aims to determine the loss of weight caused by workout intensity.

So, *Workout intensity* is the explanatory variable.

MATCHED EXAMPLE 2

a The explanatory variable is *Age group*.

The response variable is *Movie preference*.

b Age Group

Movie preference	Children	Adult	Total
Animated	19	5	24
Live-action	11	15	26
Total	30	20	50

MATCHED EXAMPLE 3

a

Movie preference	Age group	
	Children	Adult
Animated	71%	30%
Live-action	29%	70%
Total	100%	100%

b The table suggests there is an association between the age group and the type of movies preferred. 71% of children prefer movies that are animated compared to only 30% of adults. The difference in movie preference between children and adults is over 40%, indicating that children have a greater preference for animated movies than adults do.

MATCHED EXAMPLE 4

There are considerable differences in the percentages of demand for oranges and grapes in summer (oranges 36% and grapes 20%), spring (oranges 23% and grapes 40%), autumn (oranges 12% and grapes 25%) and winter (oranges 29% and grapes 15%).

The significant differences in the demand percentages for oranges and grapes in different seasons suggests that there is an association between the demand for the two fruits and the seasons.

MATCHED EXAMPLE 5

a **i** Inorganic Fertiliser range = 5.8

Organic Fertiliser range = 3.5

ii Inorganic Fertiliser median = 4.1

Organic Fertiliser median = 2.1

iii Inorganic Fertiliser IQR = 2.4

Organic Fertiliser IQR = 1.8

b **i** The inorganic fertiliser data is negatively skewed.

ii The organic fertiliser data is positively skewed.

c The organic fertiliser range (3.5 m) is considerably less than the inorganic fertiliser range (5.8 m).

The organic fertiliser median (2.1 m) is less than the inorganic fertiliser median (4.1 m). The organic fertiliser IQR (1.8 m) is considerably less than the inorganic fertiliser IQR (2.4 m). The organic fertiliser data distribution is positively skewed while the inorganic fertiliser data distribution is negatively skewed. Yes, the back-to-back stem plot supports the contention that the height of the tree is associated with the type of fertiliser.

MATCHED EXAMPLE 6

a The Quarter 1 data is symmetrical.

The Quarter 2 data is negatively skewed.

The Quarter 3 data is positively skewed.

b Quarter 1: median = 400, range = 600, IQR = 200

Quarter 2: median = 550, range = 600, IQR = 300

Quarter 3: median = 250, range = 800, IQR = 500

c Yes, the dot plots support the contention that the lolly sales of the chocolate company are associated with the period of sales. The Quarter 1 IQR (200 lollies) is noticeably less than the Quarter 2 IQR (300 lollies), which is noticeably less than the Quarter 3 IQR (500 lollies).

MATCHED EXAMPLE 7

a The location 1 data is positively skewed.

The location 2 data is negatively skewed.

The location 3 data is approximately symmetrical.

b Location 1: median = 270, range = 560, IQR = 250

Location 2: median = 450, range = 620, IQR = 240

Location 3: median = 410, range = 580, IQR = 310

c Yes, the boxplots support the contention that the number of books sold is associated with the store location. The shapes of the three boxplots are different. Store location 1 data is positively skewed, store location 2 data is negatively skewed and store location 3 data is approximately symmetrical. The range of sales of store location 2 is considerably higher than that of the other two stores.

MATCHED EXAMPLE 8

a *Age (months)*

b *Weight (kg)*

c 15 puppies

d A six-month-old puppy whose weight is 15 kg

e 7 puppies

f 10 months

MATCHED EXAMPLE 9

a **i** There is no association.

ii There appears to be no association between age and the number of siblings.

b **i** Positive, linear and strong

ii Distance increases as time increases.

9780170464055

c **i** Negative, linear and strong

    ii The number of jumpers sold tends to increase as the temperature decreases.

MATCHED EXAMPLE 10

a **i** Scatterplot

    ii Both *Batting average* and *Age* are numerical.

b **i** Parallel percentage segmented bar charts

    ii Both *Eye colours and Home states* are categorical.

c **i** Back-to-back stem plot

    ii *Cities (Sydney, Melbourne)* is categorical with two categories and *Number of books* is numerical, and seeing the data values is important.

d **i** Parallel dot plots and parallel boxplots

    ii *Schools are categorical with more than two categories and Test scores are numerical.*

MATCHED EXAMPLE 11

a The data suggests there is no association between a person's *Age* and *the Number of pets* they own.

b The data suggests there is a moderate negative linear association between the number of coffees a person drinks and the number of hours they sleep.

c The data suggests there is a strong positive linear association between the lengths of the trains and the time taken by them to cross a railway station.

MATCHED EXAMPLE 12

a *The number of customers* could be the underlying cause of the correlation between the two.

b *Climate changes* could be the underlying cause of the correlation between the two.

c The *battery capacity* could be the underlying cause of the correlation between the two.

d *Height* could be the cause of the correlation between the two.

CHAPTER 3

MATCHED EXAMPLE 1

a Explanatory variable: *balls faced*

    Response variable: *runs scored*

b $\overline{x} = 20.88$, $s_x = 16.21$,

    $\overline{y} = 10.02$, $s_y = 9.45$, $r = 0.56$

c *runs scored* $= 3.2 + 0.33 \times$ *balls faced*

MATCHED EXAMPLE 2

a The *y*-intercept is 20.

The temperature was 20°C before heating.

b Slope $= 15.0$

The temperature of the flask on an average increase by 15°C for every one-minute increase in the time of heating.

c *temperature* $= 20 + 15.0 \times$ *time*

MATCHED EXAMPLE 3

a The slope is $+2.73$.

This means that on average, the productivity increases by 2.73 units for every one-hour increase in man-hours.

b The *y*-intercept is -5.61.

This means that productivity is -5.61 units when man-hours is 0 hours. The productivity cannot be -5.61 units before the work begins. This least squares line of best fit only applies from certain minimum man-hours.

MATCHED EXAMPLE 4

a $r^2 = 0.863$

b 86.3% of the variation in the number of cars can be explained by the variation in the number of motorcycles.

c $100 - 86.3 = 13.7$

14% of variation in the number of cars is not explained by the variation in the number of motorcycles.

d *number of cars* $= 9.2 + 1.1 \times$ *number of motorcycles*

e Yes, this is an appropriate model because of the high r^2 value of 86.3%.

MATCHED EXAMPLE 5

a $r = -0.90$

b $r = 0.85$

MATCHED EXAMPLE 6

a 82.5 minutes

2 is within the original data range of 1 to 10 years, so this involves interpolation.

b 432.5 minutes

12 is outside of the original data range of 1 to 10 years, so this involves extrapolation.

c The prediction for the 2-year-old involves interpolation, so it is more reliable than the prediction for a 12-year-old, which involves extrapolation.

d *age* $= 8.5$ years

MATCHED EXAMPLE 7

a **i** Residual value for Robert $= \$500.00$

    ii Residual value for David $= -\$500.00$

    iii Residual value for Riya $= -\$3000.00$

    iv Residual value for Sam $= \$0.00$

b **i** Residual value $= 156 - 158.1 = -2.1$cm

    ii Residual value $= 180 - 166.9 = 13.1$ cm

MATCHED EXAMPLE 8

a *sales of umbrella* $= a + b \times \dfrac{1}{temperature}$

  sales of umbrella $= a + b \times \log(temperature)$

  $\dfrac{1}{sales \ of \ umbrella} = a + b \times (temperature)$

  $(sales \ of \ umbrella)^2 = a + b \times (temperature)$

b *cholesterol level* $= a + b \times (exercise)^2$

  $(cholesterol \ level)^2 = a + b \times exercise$

MATCHED EXAMPLE 9

a **i** The predicted sales is $1176 when the number of customers is 30.

  The predicted sales is $2512 when the number of customers is 30.

  ii The predicted sales is $3939 when the number of customers is 55.

  The predicted sale is $4467 when the number of customers is 55.

  iii The predicted sale is $8326 when the number of customers is 80.

  The predicted sale is $7943 when the number of customers is 80.

b

c The coefficient of determination for the squared transformation is closer to 1, so it gives the best fit to the data.

CHAPTER 4

MATCHED EXAMPLE 1

median $= 1100$ audience members

MATCHED EXAMPLE 2

a Sales of washing machines are likely to show seasonality because more people would have the time to shop for them on weekends rather than weekdays.

b Families going for vacation are likely to show seasonality because there would be a large number during holidays and few otherwise.

c Sales of summer clothes are likely to show seasonality because more would be bought in summer than in winter.

d Sales of butter are not likely to show seasonality because butter is eaten regularly each day.

MATCHED EXAMPLE 3

a

Year	Marks scored	Three-point moving mean
1	40	
2	60	$\dfrac{40+60+75}{3} = 58.33$
3	75	$\dfrac{60+75+70}{3} = 68.33$
4	70	$\dfrac{75+70+89}{3} = 78$
5	89	$\dfrac{70+89+80}{3} = 79.67$
6	80	$\dfrac{89+80+70}{3} = 79.67$
7	70	$\dfrac{80+70+82}{3} = 77.33$
8	82	$\dfrac{70+82+75}{3} = 75.67$
9	75	$\dfrac{82+75+80}{3} = 79$
10	80	$\dfrac{75+80+85}{3} = 80$
11	85	$\dfrac{80+85+95}{3} = 86.67$
12	95	

b The smoothed marks scored for the fifth year is 79.67. There was not enough data to calculate the smoothed marks scored for the twelfth year.

c

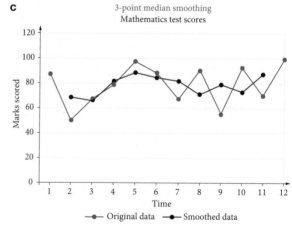

3-point median smoothing
Mathematics test scores

d The graph of the smoothed data indicates an increasing trend.

a

Month	Single-family housing permits	Four-point moving means	Four-point moving mean with centring
Jan	7540		
Feb	6005		
		$\dfrac{7540 + 6005 + 6509 + 5890}{4} = 6486$	
Mar	6509		$\dfrac{6486 + 5849}{2} = 6167$
		$\dfrac{6005 + 6509 + 5890 + 4990}{4} = 5848.5$	
Apr	5890		$\dfrac{5849 + 5472}{2} = 5660$
		$\dfrac{6509 + 5890 + 4990 + 4500}{4} = 5472$	
May	4990		$\dfrac{5472 + 5220}{2} = 5346$
		$\dfrac{5890 + 4990 + 4500 + 5500}{4} = 5220$	
Jun	4500		$\dfrac{5220 + 5048}{2} = 5134$
		$\dfrac{4990 + 4500 + 5500 + 5200}{4} = 5047.5$	
Jul	5500		$\dfrac{5048 + 5049}{2} = 5048$
		$\dfrac{4500 + 5500 + 5200 + 4995}{4} = 5049$	
Aug	5200		$\dfrac{5049 + 5018}{2} = 5034$
		$\dfrac{5500 + 5200 + 4995 + 4378}{4} = 5018$	
Sep	4995		$\dfrac{5018 + 4641}{2} = 4830$
		$\dfrac{5200 + 4995 + 4378 + 3990}{4} = 4641$	
Oct	4378		$\dfrac{4641 + 4091}{2} = 4366$
		$\dfrac{4995 + 4378 + 3990 + 3000}{4} = 4091$	
Nov	3990		
Dec	3000		

b The smoothed residential building permits for April is 5660. There is not enough data to calculate the smoothed residential building permits for November.

c

Housing permits (smoothed data)

d The graph of the smoothed data indicates a decreasing trend.

e The smoothed data value for February is 6515 and that for March is 6228.

MATCHED EXAMPLE 5

a

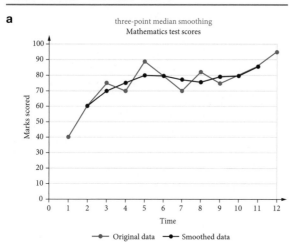

three-point median smoothing
Mathematics test scores

b

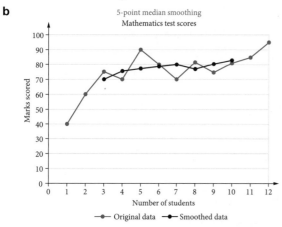

5-point median smoothing
Mathematics test scores

c The graphs of the smoothed data indicate an increasing trend.

MATCHED EXAMPLE 6

a Two days have below average sales.

b Monday

c Sunday

d 7

e 1

f

Mon	Tue	Wed	Thu	Fri	Sat	Sun
101%	130%	41%	30%	133%	105%	160%

g 100%

h 70%

MATCHED EXAMPLE 7 ANSWERS

a

Quarter 1	Quarter 2	Quarter 3	Quarter 4
$\frac{54}{37} = 1.46$	$\frac{32}{37} = 0.86$	$\frac{44}{37} = 1.19$	$\frac{18}{37} = 0.49$

b

Quarter 1	Quarter 2	Quarter 3	Quarter 4
146%	86%	119%	49%

c Quarter 4

MATCHED EXAMPLE 8

a

Year	Q1	Q2	Q3	Q4
Seasonal index	1.0016	1.2523	0.8429	0.9032

b

Year	Q1	Q2	Q3	Q4
2023	64.90	53.50	47.46	33.22
2024	44.93	63.88	59.32	71.97
2025	74.88	67.88	83.05	88.57

c

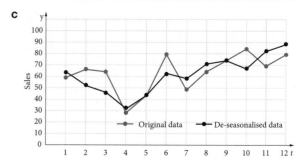

MATCHED EXAMPLE 9

a The quarter 4 seasonal index is 0.9.

b 11%.

c $4 500 000

MATCHED EXAMPLE 10

a *de-seasonalised number of sales* = 2.18 + 3.49 × *quarter number*

b

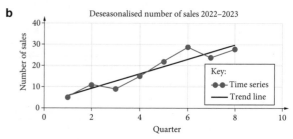

Deseasonalised number of sales 2022–2023

During 2022–2023, the sales of the ring increased on average by 3.49 per quarter.

c De-seasonalised number forecast to be sold in Q2 2024 is 37.

d Actual number forecast to be sold in Q3 2024 is 33.

CHAPTER 5

MATCHED EXAMPLE 1

a Start with 1. Multiply each value by 2 and then add 2 to find the next value.

b 1, 4, 10, 22

c $u_4 = 46$

$u_5 = 94$

MATCHED EXAMPLE 2

a $u_0 = 2$

$u_{n+1} = \dfrac{1}{2} u_n - 0.25$

b $u_0 = 6, u_{n+1} = 5u_n$

c $u_0 = -10, u_{n+1} = u_{n-5}$

MATCHED EXAMPLE 3

a True

b False

c False

d False

MATCHED EXAMPLE 4

a $80

b **i** Oliver's bank account balance after four years is $4320.

   ii Oliver's balance is first greater than $4300 after four years.

   iii Total amount of interest earned after six years is $480.

c $V_0 = 4000, V_{n+1} = V_n + 80$

d The recurrence relation models linear growth.

e

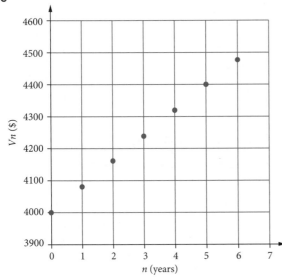

MATCHED EXAMPLE 5

a The balance after six years is $5900.

b The balance after six years is $4100.

MATCHED EXAMPLE 6

a The fixed amount of interest paid each year is $500.

b $V_n = 10\,000 + 500n$

c The balance of Ellyse's account after eight years is $14\,000.

d It would take 20 years for the investment to double.

MATCHED EXAMPLE 7

a $310

b **i** The value of the mobile phone after four years is $310.

   ii The value of the mobile phone first falls below $700 after three years.

   iii The mobile phone depreciates to zero after five years.

c $V_0 = 1550, V_{n+1} = V_n - 310$

d The recurrence relation models linear decay.

e

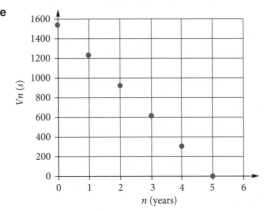

MATCHED EXAMPLE 8

a The fixed amount of depreciation paid each year is $90\,000.

b $V_n = 900\,000 - 90\,000n$

c The value of the boat after six years is $360\,000.

d It would take 10 years for the boat to depreciate to zero.

MATCHED EXAMPLE 9

a The fixed amount of depreciation each year is $450.

$V_n = V_0 - nd$

b The annual flat rate of depreciation is 15%.

MATCHED EXAMPLE 10

a The amount of depreciation is determined by applying a rate per unit of use—$20\,000 every time the game is played.

b The value of the game after four plays is $120\,000.

The value of the game first falls below $120\,000 after five plays.

c $V_0 = 200\,000, V_{n+1} = V_n - 20\,000$

d The recurrence relation models linear decay.

e

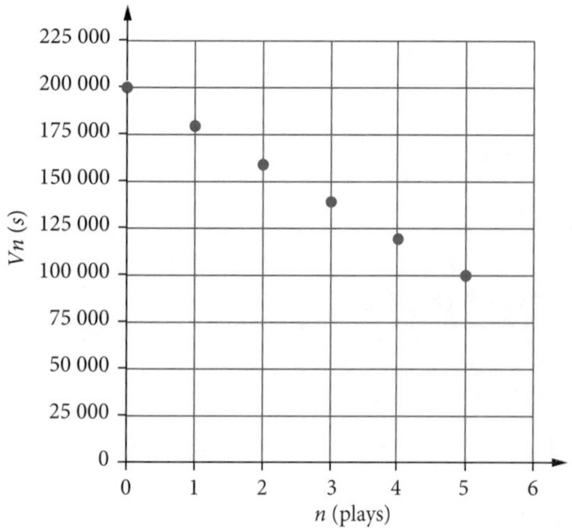

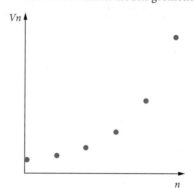

Graph with y-axis labeled V_n ($) ranging 0 to 225 000, x-axis labeled n (plays) from 0 to 6. Points plotted at approximately (0, 200 000), (1, 180 000), (2, 160 000), (3, 140 000), (4, 120 000), (5, 100 000).

iii The percentage interest rate per day = $\dfrac{5}{12}$ %.

iv The amount of interest earned on the first month is $333.33.

MATCHED EXAMPLE 15

a $V_0 = 5000$, $V_{n+1} = 1.04V_n$

b $V_0 = 8000$, $V_{n+1} = 1.003V_n$

c $V_0 = 10\,000$, $V_{n+1} = 1.00125V_n$

d The recurrence relation models geometric growth.

e

Graph with axis labeled V_n vertical and n horizontal. Points plotted showing increasing growth.

MATCHED EXAMPLE 11

a $V_n = 70\,000 - 0.80n$

b The value of the concrete mixer after it has mixed a total of 60 000 kilograms of concrete is $22 000.

c The concrete mixer has mixed 50 000 kg when it depreciates to $30 000.

d The concrete mixer has mixed 87 500 kg when it depreciates to zero.

MATCHED EXAMPLE 12

a The depreciation in the value of the water bottle labelling machine is $0.05 per label.

b $V_n = 12\,000 - 0.05n$

MATCHED EXAMPLE 16

a Grace initially invested $10 000.

b $V_3 = 1.05V_2 = 1.05 \times 11\,025 = 11\,576.25$

c The annual percentage compound interest rate is 5%.

d The balance of Grace's account first exceeds $13 000 after six years.

e Total interest = $3400.96

MATCHED EXAMPLE 13

a

| | Compound | | Simple | |
n	Interest ($)	Value of investment ($)	Interest ($)	Value of investment ($)
3	$\dfrac{4}{100} \times 4326.40 = 173.06$	$4326.40 + 173.06 = 4499.46$	$\dfrac{4}{100} \times 4000 = 160$	$4320 + 160 = 4480$
4	$\dfrac{4}{100} \times 4499.46 = 179.98$	$4499.46 + 179.98 = 4679.44$	$\dfrac{4}{100} \times 4000 = 160$	$4480 + 160 = 4640$
5	$\dfrac{4}{100} \times 4679.44 = 187.18$	$4679.44 + 187.18 = 4866.62$	$\dfrac{4}{100} \times 4000 = 160$	$4640 + 160 = 4800$

b The value of the compound interest investment after five years is $4866.62.

c The compound interest investment has $66.62 more.

MATCHED EXAMPLE 14

a **i** There are four quarterly compounding periods per year.

ii There are $4 \times 10 = 40$ quarterly compounding periods over 10 years.

iii The percentage interest rate per quarter = $\dfrac{10}{4}$% .

iv The amount of interest earned in the first quarter is $250.

b **i** There are 12 monthly compounding periods per year.

ii There are $12 \times 10 = 120$ monthly compounding periods over 10 years.

MATCHED EXAMPLE 17

a The percentage interest rate per month = $\dfrac{3.6}{12}$ % = 0.3%.

b $V_n = 1.003^n \times 25\,000$

c The value of the investment after five years is $29 922.37

d The value of the investment after five years is $29 906.34.

e Monthly compounding gives a larger balance than quarterly compounding after five years.

MATCHED EXAMPLE 18

a $r = 6.2\%$

b It will take 24 quarters for the investment to grow to $10 000.

c $15 000

MATCHED EXAMPLE 19

a Bank 1

$r_{effective} = 5.65\%$ p.a.

Bank 2

$r_{effective} = 5.59\%$ p.a.

Bank 3

$r_{effective} = 5.88\%$ p.a.

Bank 4

$r_{effective} = 5.85\%$ p.a.

b Valerie should choose Bank 3 because it pays the highest effective rate of interest and will therefore pay more interest.

c Bank 2 would earn Valerie the least interest.

d The nominal and effective interest rates for Bank 4 are the same because the rate compounds annually.

MATCHED EXAMPLE 20

a **i** The value of the truck after six years is $17 039.

ii The amount of depreciation in the fourth year is $6656.

iii The truck first depreciates to under $25 000 after five years.

b $V_0 = 65\,000, V_{n+1} = 0.80V_n$

c The value of the truck in any year is 80% of its value the previous year.

d The recurrence relation models geometric decay.

e The value of the truck after eight years is around $11 000.

MATCHED EXAMPLE 21

a $V_2 = 486$

b The annual percentage rate of depreciation is 10%.

c The sewing machine will be sold after seven years.

MATCHED EXAMPLE 22

a $V_n = 0.9^n \times 2600$

b The value of the refrigerator after 11 years is $816.

c It will take the refrigerator 16 years to depreciate to under $500.

d The amount of depreciation in the fourth year is $190.

MATCHED EXAMPLE 23

a $r = 10\%$

b The original price of the bulldozer is $395 000.

CHAPTER 6

MATCHED EXAMPLE 1

a $118 620.17

b Seven years

c **i** 74 months

ii 10 months

MATCHED EXAMPLE 2

a 12.3%

b $602

MATCHED EXAMPLE 3

a The amount of money that Steve initially invested was $9928.

b $V_4 = \$8707.13$

MATCHED EXAMPLE 4

a $V_0 = 25\,000, V_{n+1} = 1.026\,25V_n - 1580$

b The recurrence relation models a combination of linear decay *and* geometric growth.

c

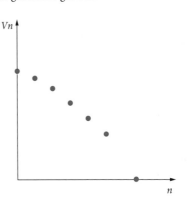

MATCHED EXAMPLE 5

a $8000

b $250

c $V_0 = 8000$

$V_1 = 1.0085V_0 - 250 = 7818.00$

$V_2 = 1.0085V_1 - 250 = 7634.45$

$V_3 = 1.0085V_2 - 250 = 7449$

d 10.2%

e Seven months

MATCHED EXAMPLE 6

a $r = \dfrac{7068.00}{310\,000.00} \times 100 = 0.0228 \times 100 = 2.28\%$

$r = \dfrac{7018.93}{307\,848.00} \times 100 = 0.0228 \times 100 = 2.28\%$

b 9.12% per annum compounding monthly.

c

Payment number	Payment	Interest	Principal reduction	Balance
0	0.00	0.00	0.00	310 000.00
1	9220.00	7068.00	2152.00	307 848.00
2	9220.00	7018.93	2201.07	305 646.93
3	9220.00	6968.75	2251.25	303 395.68
4	9220.00	6917.42	2302.58	301 093.11

MATCHED EXAMPLE 7

a $30 888.15

b $30 886.60

c $3015.20

d $84 111.85 of the principal has been paid after three payments.

e The loan has cost Monica $14 725.56.

f The balance after the last payment is $1.55. It should be $0 if the loan has been fully paid out.

g $32 432.55

MATCHED EXAMPLE 8

a $V_1 = 20 250$

$V_2 = 20 250$

The value of the loan stays at the principal value of $20 250 for all compounding periods.

b $100 000

c $230

MATCHED EXAMPLE 9

The annual interest rate is 8.7%.

The value is negative because Patrick is making payments to the bank, so the money is moving *away* from him.

MATCHED EXAMPLE 10

a Gimli's quarterly repayment is $1631.02.

b $52 192.64

c $12 192.64

MATCHED EXAMPLE 11

a **$138 952.68**

b 10%

c James will need to make 68 repayments.

d James still owes $88.76.

e $4088.76

MATCHED EXAMPLE 12

a The annual interest rate is 6.2%.

b Mia's monthly repayments are $120.83.

MATCHED EXAMPLE 13

It will take $13\frac{1}{2}$ years for Henry to repay the loan in full.

MATCHED EXAMPLE 14

a $1858.18

b $76 946.96

MATCHED EXAMPLE 15

a $175   **b** $931.30

c $1483.36

MATCHED EXAMPLE 16

a $V_0 = 135 000$, $V_{n+1} = 1.0085 V_n - 150$

b 15.6%

c The investment increased by $257.30.

MATCHED EXAMPLE 17 ANSWERS

a $500 000.00   **b** $10 250.00

c 1.8%   **d** 7.2%

e $8977.50   **f** $1272.50

g $497 477.50

MATCHED EXAMPLE 18

The annuity will last 16 years.

MATCHED EXAMPLE 19

The amount of each monthly payment is $4050.98.

Payment number	Payment	Interest	Principal addition	Balance
0	0.00	0.00	0.00	6200.00
1	600.00	620.00	1220.00	7420.00
2	600.00	742.00	1342.00	8762.00
3	600.00	876.20	1476.20	10 238.20
4	600.00	1023.82	1623.82	11 862.02

MATCHED EXAMPLE 20

a $V_1 = 75 000$

$V_2 = 75 000$

The value of the investment stays at the principal value $75 000 for all compounding periods.

b Michael should invest $40 000.

c Peter will receive $3150.00 quarterly from this investment.

MATCHED EXAMPLE 21

a The annual interest rate is 5.3%.

b Peter will receive $3150.00 quarterly from this investment.

MATCHED EXAMPLE 22

a $V_0 = 85 000$, $V_{n+1} = 1.013 V_n + 550$

b 4.2%

c The investment increased by $856.43.

MATCHED EXAMPLE 23

a $r = \dfrac{620.00}{6200.00} \times 100 = 0.1 \times 100 = 10\%$

$r = \dfrac{742.00}{7420.00} \times 100 = 0.1 \times 100 = 10\%$

b 10% per annum compounding annually

c

Payment number	Payment	Interest	Principal addition	Balance
0	0.00	0.00	0.00	6200.00
1	600.00	620.00	1220.00	7420.00
2	600.00	742.00	1342.00	8762.00
3	600.00	876.20	1476.20	10 238.20
4	600.00	1023.82	1623.82	11 862.02

MATCHED EXAMPLE 24

a 2%

b $189.40

c $989.40

d $880.00

MATCHED EXAMPLE 25

a The value of the investment will first exceed $85 000 after eight quarters.

b Alex increased his payments to $852.66.

MATCHED EXAMPLE 26

Addison has **$887 962.37** in her account after she has been retired for eight years.

CHAPTER 7

MATCHED EXAMPLE 1

a $A = \begin{bmatrix} 50 & 20 & 45 & 20 \\ 100 & 50 & 35 & 10 \\ 120 & 60 & 65 & 15 \\ 95 & 35 & 70 & 30 \end{bmatrix}$

The order of A is 4×4.

A has 16 elements.

b [35] The order is 1×1.

c $\begin{bmatrix} 120 & 60 & 65 & 15 \end{bmatrix}$

d $\begin{bmatrix} 50 & 100 & 120 & 95 \end{bmatrix}$

e $\begin{bmatrix} 50 & 45 \\ 120 & 65 \end{bmatrix}$

f $\begin{bmatrix} 135 \\ 195 \\ 260 \\ 230 \end{bmatrix}$

g

	S	P	F	C
Supermarket	50	100	120	95
Theatre	20	50	60	35
Petrol station	45	35	65	70
Pharmacy	20	10	15	30

MATCHED EXAMPLE 2

a 3×3, binary matrix, square matrix

b 4×4, square matrix, binary matrix, zero matrix

c 4×4, square matrix, binary matrix, permutation matrix

d 5×1, column matrix, binary matrix, summing matrix

e 3×3, square matrix

MATCHED EXAMPLE 3

a $A^T = \begin{bmatrix} 4 & 0 & 2 & 3 \\ 3 & 0 & -1 & 2 \\ 1 & 1 & 2 & 4 \end{bmatrix}$

The order of A is 4×3. The order of A^T is 3×4.

b $A^T = \begin{bmatrix} 1 & 2 & 4 \\ 1 & 3 & 0 \end{bmatrix}$

The order of A is 3×2. The order of A^T is 2×3.

c $A^T = \begin{bmatrix} 1 & 2 & 0 & 3 \end{bmatrix}$

The order of A is 4×1. The order of A^T is 1×4.

MATCHED EXAMPLE 4

a This is not a square matrix, so it has no leading diagonal and cannot be any of the options.

b 1, 2, 3,0 upper triangular matrix

c 1, 0, 3, 2, −1 symmetric matrix

d 1, 1, 1 lower triangular matrix

e 1, 1, 1 upper triangular matrix

MATCHED EXAMPLE 5

a m_{22} tells us that 12 nectarines were sold on Tuesday.

b m_{34} tells us that 16 apricots were sold on Wednesday.

c $m_{11} + m_{12} + m_{13}$ tells us that a total of $20 + 15 + 17 + 11 = 63$ oranges were sold on the first four days of the week.

d $m_{41} + m_{42} + m_{43} + m_{44}$ tells us that a total of $11 + 16 + 12 + 15 = 54$ fruits were sold on Thursday.

MATCHED EXAMPLE 6

a $A = \begin{bmatrix} 1 & 2 & 3 \\ -2 & 1 & 0 \\ -3 & 0 & 1 \end{bmatrix}$

b $A = \begin{bmatrix} 1 & 1 \\ 2 & 2 \\ 3 & 3 \\ 4 & 4 \end{bmatrix}$

c $A = \begin{bmatrix} 5 & 0 \\ 0 & 7 \end{bmatrix}$

d $A = \begin{bmatrix} 1 & 0 \\ 1 & 1 \\ 1 & 1 \end{bmatrix}$

MATCHED EXAMPLE 7

a $\begin{bmatrix} 4 & 8 \\ 0 & -4 \\ 0 & 0 \end{bmatrix}$

b $\begin{bmatrix} -2 & -1 & 0 \end{bmatrix}$

c $\begin{bmatrix} 3 & 2 \\ 0 & 0 \\ 4 & 2 \end{bmatrix}$

d $3C - B$ is not defined because the order of $3C$ is 1×3 and the order of B is 3×2. Matrices must have the same order for subtraction to be possible.

e $\begin{bmatrix} -\dfrac{1}{2} & -\dfrac{1}{2} & 1 \end{bmatrix}$

f $\begin{bmatrix} 0.25 & 0.5 \\ 0 & -0.25 \\ 0 & 0 \end{bmatrix}$

MATCHED EXAMPLE 8

a $x = 12, y = 2$

b $A = \begin{bmatrix} -1 & 12 \\ 5 & 4 \end{bmatrix}$

c $2A + B = \begin{bmatrix} 4 & 7 & 8 \\ 9 & 8 & 11 \\ 12 & 13 & 12 \end{bmatrix}$

MATCHED EXAMPLE 9

a $\dfrac{1}{2} \times \begin{bmatrix} 48 \\ 40 \\ 42 \\ 44 \end{bmatrix}$

b $\dfrac{1}{2} \times \begin{bmatrix} 48 \\ 40 \\ 42 \\ 44 \end{bmatrix} - \begin{bmatrix} 2 \\ 2 \\ 2 \\ 2 \end{bmatrix}$

c $\dfrac{1}{2} \times \begin{bmatrix} 48 \\ 40 \\ 42 \\ 44 \end{bmatrix} - \begin{bmatrix} 2 \\ 2 \\ 2 \\ 2 \end{bmatrix} + \begin{bmatrix} 10 \\ 10 \\ 10 \\ 10 \end{bmatrix}$

MATCHED EXAMPLE 10

a **i** MO is not defined.

b **i** NO is defined.

   ii NO has order 3×2.

   iii $NO = \begin{bmatrix} 9 & 5 \\ 2 & 1 \\ -1 & 1 \end{bmatrix}$

c **i** MN is not defined.

d **i** MP is defined.

   ii MP has order 3×3.

   iii $MP = \begin{bmatrix} 4 & -2 & 0 \\ 0 & 0 & 0 \\ 2 & -1 & 0 \end{bmatrix}$

e **i** N^4 is not defined.

f **i** PM is defined.

   PM has order 1×1.

   $PM = [4]$

g **i** $O^2 - O$ is defined.

   ii $O^2 - O$ has order 2×2.

   iii $O^2 - O = \begin{bmatrix} 3 & 2 \\ 2 & 1 \end{bmatrix}$

MATCHED EXAMPLE 11

a **i** $\begin{bmatrix} 1 & 2 & 0 \\ 1 & 1 & 0 \\ 2 & 0 & 1 \end{bmatrix} \begin{bmatrix} 3 \\ x \\ 1 \end{bmatrix} = \begin{bmatrix} (1\times3) + (2\times x) + (0\times1) \\ (1\times3) + (1\times x) + (0\times1) \\ (2\times3) + (0\times x) + (1\times1) \end{bmatrix}$

$= \begin{bmatrix} 3+2x \\ 3+x \\ 7 \end{bmatrix}$

$\begin{bmatrix} 3 \\ y \\ 7 \end{bmatrix} + \begin{bmatrix} y \\ 1 \\ 0 \end{bmatrix} = \begin{bmatrix} 3+y \\ y+1 \\ 7 \end{bmatrix}$

   ii If $\begin{bmatrix} 3+2x \\ 3+x \\ 7 \end{bmatrix} = \begin{bmatrix} 3+y \\ y+1 \\ 7 \end{bmatrix}$, then

$\begin{array}{ll} 3+2x=3+y & 3+x=y+1 \\ y=2+3-3 & y=3+x-1 \\ y=2x & y=2+x \end{array}$

b **i** Let M have order $m \times m$.

For MN to be defined N needs to have order $m \times 1$.

For NO to be defined O needs to have order $1 \times m$.

ii MNO has order

$\begin{array}{ccc} M & N & O \\ (m \times m) & (m \times 1) & (1 \times m) \end{array}$

$\begin{array}{cc} MN & O \\ (m \times 1) & (1 \times m) \end{array}$

$\begin{array}{c} MNO \\ (m \times m) \end{array}$

So, MNO is a square matrix.

MATCHED EXAMPLE 12

a $\begin{bmatrix} 2 \\ 3 \\ 4 \\ 3 \end{bmatrix}$ **b** $\begin{bmatrix} 10 & 7 \\ 4 & 4 \end{bmatrix}$

MATCHED EXAMPLE 13

a $\begin{bmatrix} S \\ I \\ L \\ E \\ N \\ T \end{bmatrix}$ **b** $P = \begin{bmatrix} 0 & 0 & 0 & 0 & 1 \\ 0 & 1 & 0 & 0 & 0 \\ 0 & 0 & 1 & 0 & 0 \\ 0 & 0 & 0 & 1 & 0 \\ 1 & 0 & 0 & 0 & 0 \end{bmatrix}$

MATCHED EXAMPLE 14

There is no answer key for this question.

MATCHED EXAMPLE 15

a **i** $\det(A) = 0$

   ii Since $\det(A) = 0$, A^{-1} does not exist.

b **i** $\det(B) = 4$

   ii $B^{-1} = \begin{bmatrix} \dfrac{1}{2} & -3 \\ -\dfrac{1}{4} & 2 \end{bmatrix}$

c **i** $\det(C) = -3$

   ii $C^{-1} = \begin{bmatrix} -1 & 1 \\ \dfrac{4}{3} & -1 \end{bmatrix}$

MATCHED EXAMPLE 16

a $a = 3$ **b** $a = 1$

MATCHED EXAMPLE 17

a $P = \begin{bmatrix} 3 \\ 2 \\ 1 \end{bmatrix}$ **b** $M = \begin{bmatrix} 1 & 1 & 0 \\ 0 & 1 & 1 \\ 0 & 0 & 0 \\ 2 & 0 & 2 \\ 0 & 1 & 0 \\ 0 & 0 & 2 \\ 0 & 0 & 1 \end{bmatrix}$

c $\begin{bmatrix} 5 \\ 3 \\ 0 \\ 8 \\ 2 \\ 2 \\ 1 \end{bmatrix}$

9780170464055

d Alex won with 8 points.

e 21 is less than 28, so Melbourne Little Champions lost.

MATCHED EXAMPLE 18

a i $P = \begin{bmatrix} 11 \\ 14 \\ 10 \\ 17 \end{bmatrix}$

ii p_{31} tells us that the total number of T-shirts made by Amy is 10.

b i $Q = \begin{bmatrix} 15 & 14 & 14 & 9 \end{bmatrix}$

Q gives the total number of each colour of T-shirts made by the designing class members.

ii q_{14} tells us that a total of nine pink T-shirts were made by the members of the designing class.

c $[13]$

$\dfrac{1}{4} NTM$ tells us that the mean number of T-shirts made by the members of the T-shirt designing class is 13.

MATCHED EXAMPLE 19

a $\begin{bmatrix} 3800 \\ 2410 \end{bmatrix}$

b i $\begin{bmatrix} 4500 & 6000 \\ 2300 & 1200 \\ 3800 & 2410 \\ 1000 & 3250 \end{bmatrix}$

ii $\begin{bmatrix} 6300.00 & 8400.00 \\ 3220.00 & 1680.00 \\ 5320.00 & 3374.00 \\ 1400.00 & 4550.00 \end{bmatrix}$

c i $\begin{bmatrix} 1800 & 2400 \\ 920 & 480 \\ 1520 & 964 \\ 400 & 1300 \end{bmatrix}$

ii $9784.00

MATCHED EXAMPLE 20

a i Zoe can send direct messages to Emma and Alexis.

Chloe can send direct messages to Zoe and Alexis.

Emma can send direct messages to Chloe and Alexis.

Alexis can send direct messages to Emma.

ii The leading diagonal represents links where the sender and receiver are the same. This is not considered communication, so they are redundant links.

iii

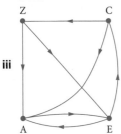

iv Zoe → Emma → Chloe

b i

receiver

		A	B	C	D	O
	A	0	0	0	1	1
	B	0	0	1	0	1
sender	C	0	1	0	0	0
	D	1	0	0	0	0
	O	1	1	0	0	0

ii The matrix is symmetric because all the communications go both ways.

MATCHED EXAMPLE 21

a There are two ways Q can connect with R by connecting directly to one other computer.

b $Q \to P \to R$ and $Q \to S \to R$

c The total number of redundant two-step connections is 8.

d There are two redundant two-step connections from P to P:

$P \to Q \to P$ and $P \to R \to P$

e The total number of one-step and two-step connections from S to R is 1.

MATCHED EXAMPLE 22

a i

Loser

		A	B	C	D	E
	A	0	1	1	0	1
	B	0	0	1	0	1
$M = $ *Winner*	C	0	0	0	1	0
	D	1	1	0	0	1
	E	0	0	1	0	0

Loser

		A	B	C	D	E
	A	0	1	3	1	2
	B	0	0	2	1	1
$M + M^2 = $ *Winner*	C	1	1	0	1	1
	D	1	2	3	0	3
	E	0	0	1	1	0

ii

Total scores

7
4
4
9
2

iii Ranking: David (1), Alex (2), Bethany (3), Chen (3), Ella (5)

The overall chess tournament winner is David.

Loser

		P	Q	R	S	T
	P	0	0	1	1	0
	Q	1	0	1	1	1
b i $M = $ *Winner*	R	0	0	0	1	1
	S	1	0	0	0	1
	T	0	0	0	0	0

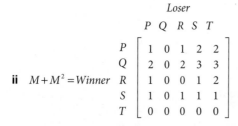

Loser

$$\begin{array}{c} & P\ \ Q\ \ R\ \ S\ \ T \\ \textbf{ii}\quad M+M^2=\text{Winner}\quad \begin{matrix} P \\ Q \\ R \\ S \\ T \end{matrix} & \begin{bmatrix} 1 & 0 & 1 & 2 & 2 \\ 2 & 0 & 2 & 3 & 3 \\ 1 & 0 & 0 & 1 & 2 \\ 1 & 0 & 1 & 1 & 1 \\ 0 & 0 & 0 & 0 & 0 \end{bmatrix} \end{array}$$

Total scores

6
10
4
4
0

iii Ranking: Quinn (1), Pearl (2), Robin (3), Sarah (3), Tina (5)

The overall winner is Quinn.

MATCHED EXAMPLE 23

Loser

$$\begin{array}{c} & P\ \ Q\ \ R\ \ S \\ \textbf{a}\quad M=\text{Winner}\quad \begin{matrix} P \\ Q \\ R \\ S \end{matrix} & \begin{bmatrix} 0 & 0 & 1 & 0 \\ 1 & 0 & 1 & 0 \\ 0 & 0 & 0 & 1 \\ 1 & 1 & 0 & 0 \end{bmatrix} \end{array}$$

b Q would be declared the clear overall winner if the result of the game between P and S was reversed.

CHAPTER 8

MATCHED EXAMPLE 1

a **i** $p = 20\%$

$q = 25\%$

$r = 55\%$

$s = 46\%$

This City

$$\begin{array}{c} & B\quad\ \ V\quad\ \ M\quad\ \ C \\ \textbf{ii}\quad \begin{bmatrix} 0.40 & 0.60 & 0.00 & 0.25 \\ 0.25 & 0.10 & 0.22 & 0.00 \\ 0.10 & 0.10 & 0.32 & 0.20 \\ 0.25 & 0.20 & 0.46 & 0.55 \end{bmatrix} & \begin{matrix} B \\ V \\ M \\ C \end{matrix}\ \ \text{Next City} \end{array}$$

This game

$$\textbf{b}\quad T = \begin{bmatrix} 0.25 & 0.40 \\ 0.75 & 0.60 \end{bmatrix}\begin{matrix} W \\ L \end{matrix}\ \ \text{Next game}$$

MATCHED EXAMPLE 2

a $a = 0.17$ $c = 0.0$ $b = 0.0$

This month

$$\begin{array}{c} P\quad\ \ \ Q\quad\ \ \ R \\ \begin{bmatrix} 0.17 & 0.70 & 0.48 \\ 0.63 & 0.30 & 0.00 \\ 0.20 & 0.00 & 0.52 \end{bmatrix}\begin{matrix} P \\ Q\ \ \text{Next month} \\ R \end{matrix} \end{array}$$

b

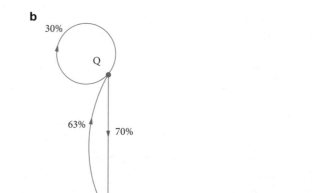

MATCHED EXAMPLE 3

a 425 Australian pelicans

b 714 Australian pelicans

c 2894 Australian pelicans are expected to be in location R the next season.

d 18.2%

MATCHED EXAMPLE 4

a P is a transition matrix because it is a square matrix, and each column adds up to 1.

b The '1' in column G and row Y means that Alex will wear the yellow shirt if he wore the green shirt the previous day.

c Alex selected the yellow shirt on the fourth day.

MATCHED EXAMPLE 5 ANSWERS

a $S_0 = \begin{bmatrix} 450 \\ 320 \end{bmatrix}$, $S_{n+1} = \begin{bmatrix} 0.90 & 0.65 \\ 0.10 & 0.35 \end{bmatrix} S_n$

b In the next week, 613 people are predicted to buy Loopy Loops and 157 people are predicted to buy Crunchy Flakes.

c Loopy Loops: $0.90 \times 450 + 0.65 \times 320 = 613$

Crunchy Flakes: $0.10 \times 450 + 0.35 \times 320 = 157$

d In the next to the next week, 654 people are predicted to buy Loopy Loops and 116 people are predicted to buy Crunchy Flakes.

MATCHED EXAMPLE 6

a Once the fish move to pond R, they do not move to the other ponds day after day.

b 5500   **c** 41%

MATCHED EXAMPLE 7

$$T = \begin{bmatrix} 0.2 & 0.5 & 0.4 \\ 0.3 & 0.2 & 0.5 \\ 0.5 & 0.3 & 0.1 \end{bmatrix}$$

MATCHED EXAMPLE 8

a After three months, 1889 ships will be at port Q, 1611 ships will be at port R and 2350 ships will be at port S.

b After eight months, 1888 ships will be at port Q, 1610 ships will be at port R and 2352 ships will be at port S.

c After 17 months, 2333 ships will be at port Q, 1794 ships will be at port R and 1723 ships will be at port S.

MATCHED EXAMPLE 9

a Since T has no zero elements, there will be an equilibrium state matrix.

b $\begin{bmatrix} 1888 \\ 1610 \\ 2352 \end{bmatrix} \begin{matrix} Q \\ R \\ S \end{matrix}$ is the equilibrium state matrix.

c In the long term, 1888 ships will be at port Q, 1610 ships will be at port R and 2352 ships will be at port S.

d 32%

MATCHED EXAMPLE 10

a Since T^2 has no zero elements, there will be an equilibrium state matrix.

b The percentage of students expected to use the internet at home each day in the long term is $0.2742 \times 100\% = 27\%$.

c In the long term, 129 students will use the internet from work to do their homework assignments.

d Expected students to use internet from school $= 1612 \times 0.4032 = 650$

e In the long term, all students end up using the internet from work to do their homework assignments.

MATCHED EXAMPLE 11

a There are 80 members with an Expert title after the first assessment.

b There are 116 members with a Master title after the second assessment.

MATCHED EXAMPLE 12

Each year, the farmer will sell 150 young quails and 1138 adult quails.

MATCHED EXAMPLE 13

a $B = \begin{bmatrix} 247 \\ -62 \\ 60 \\ 115 \end{bmatrix} \begin{matrix} C \\ V \\ S \\ L \end{matrix}$

b 62 customers who chose venison will be removed from the analysis.

c $\begin{bmatrix} 929 \\ 305 \\ 420 \\ 621 \end{bmatrix} \begin{matrix} C \\ V \\ S \\ L \end{matrix}$

MATCHED EXAMPLE 14

a The birth rates given are for baby squirrels, not just female baby squirrels. So, we need to halve these figures.

b

Age (years)	0–<2	2–<4	4–<6
Initial number	40	30	20
Birth rate	2.7	5.1	3.1
Survival rate	0.40	0.60	0

c 18 two-year-old female squirrels are expected to survive to be four.

d 323 female squirrels are expected to be born after two years.

e $L = \begin{bmatrix} 2.7 & 5.1 & 3.1 \\ 0.4 & 0 & 0 \\ 0 & 0.6 & 0 \end{bmatrix}$

MATCHED EXAMPLE 15

a The first element in the first row is zero. This means the rodents up to one year old have a birth rate of zero.

b 200 rodents

c The total number of rodents after two years is 12 640.

d 84 rodents

MATCHED EXAMPLE 16

a Increase

b The population will cycle every three time periods.

c Decrease

MATCHED EXAMPLE 17

a The long-term growth rate per time period is 0.95.

b The percentage decrease per time period is 5%.

CHAPTER 9

MATCHED EXAMPLE 1

a These two graphs are *not* isomorphic because, although they have the same number of vertices and edges, not all of the connections are the same. In the first graph, A and C are not connected, but in the second graph, they are connected.

b These two graphs are *not* isomorphic because they have different numbers of vertices and edges.

c These two graphs are isomorphic because they show exactly the same connections.

MATCHED EXAMPLE 2

a **i** 7 vertices: L, M, N, O, P, Q, R

11 edges: $LM, LR, MR, MP, MN, NP, NO, OQ, QR, PR, PQ$

6 faces: 5 enclosed and 1 outside the graph

ii

Vertex	L	M	N	O	P	Q	R	Sum
Degree	2	4	3	2	4	3	4	22

degree sum = 2 × number of edges

$$= 2 \times 11$$

$$= 22.$$

b **i** 8 vertices: *L, M, N, O, P, Q, R, S*

14 edges: *LM, LQ, LR, MS, QS, QR, RS, MN, NO* × 3, *SN, MO, MM*

8 faces: 7 enclosed and 1 outside the graph

ii

Vertex	*L*	*M*	*N*	*O*	*P*	*Q*	*R*	*S*	Sum
Degree	3	6	5	4	0	3	3	4	28

degree sum = 2 × number of edges

$$= 2 \times 14$$

$$= 28.$$

MATCHED EXAMPLE 3

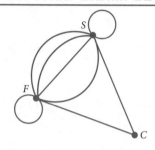

MATCHED EXAMPLE 4

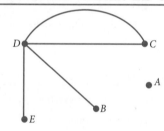

MATCHED EXAMPLE 5

a Four bridges

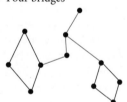

b Six bridges

c No bridges

d Two bridges

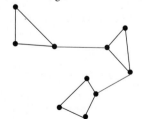

MATCHED EXAMPLE 6

a

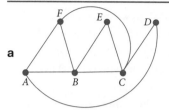

b There is a path from each vertex to every other vertex, so it is a connected graph.

c Euler's formula works for this graph.

MATCHED EXAMPLE 7

a The number of vertices is 5.

b $e = 19$

MATCHED EXAMPLE 8

a **i** It is not a simple graph because it has multiple edges.

    ii It is not a complete graph because it is not a simple graph.

    iii It is not a subgraph because the larger graph has only one edge connecting *D* and *E*.

b **i** It is a simple graph because it has no loops or multiple edges.

    ii It is a complete graph because every vertex is connected by an edge to every other vertex.

    iii It is a subgraph because it only has vertices and edges from the larger graph.

c **i** It is a simple graph because it has no loops or multiple edges.

    ii It is not a complete graph because not every vertex is connected by an edge to every other vertex.

    iii It is a subgraph because it only has vertices and edges from the larger graph.

d **i** It is not a simple graph because it has a loop.

    ii It is not a complete graph because it is not a simple graph.

    iii It is a subgraph because it only has vertices and edges from the larger graph.

MATCHED EXAMPLE 9

a This walk has no repeated edges, a repeated vertex and does not start and finish at the same vertex, so it's a trail.

b This walk has no repeated edges, a repeated vertex and starts and finish at the same vertex, so it's a circuit.

c This walk has no repeated edges, no repeated vertices and does not start and finish at the same vertex, so it's a path.

d This walk has two repeated edges, so it's a walk only.

e This walk has no repeated edges, no repeated vertices (except the first and last vertex) and starts and finishes at the same vertex, so it's a cycle.

MATCHED EXAMPLE 10

a This walk has no repeated edges, no repeated vertices (except the first and last vertex) and starts and finishes at the same vertex, so it's a cycle.

b This walk has no repeated edges, no repeated vertices and does not start and finish at the same vertex, so it's a path.

c This walk has a repeated edge *GH*, so it's a walk only.

d This walk has no repeated edges, a repeated vertex *B* and does not start and finish at the same vertex, so it's a trail.

e This walk has a repeated edge (*JH* is the same edge as *HJ*), so it's a walk only.

f This walk has no repeated edges, a repeated vertex *G* and starts and finish at the same vertex, so it's a circuit.

MATCHED EXAMPLE 11

a **i** There are 6 vertices with odd degrees:
A, B, C, D, E, F.

An Eulerian trail exists only if there are exactly two odd vertices and an Eulerian circuit only exists if all the vertices are even, so neither exists for this graph.

ii All the vertices have an even degree, so an Eulerian circuit exists.

Three Eulerian circuits are

A–B–C–D–E–C–F–A

A–F–C–E–D–C–B–A

A–B–C–E–D–C–F–A

Other answers are possible.

iii Vertex *A* and vertex *C* are the only two vertices of odd degree. Since *exactly two* vertices are of odd degree, a Eulerian trail exists.

Three Eulerian trails are

A–C–D–A–B–C

A–C–B–A–D–C

A–B–C–A–D–C

Other answers are possible.

b **i** Hamiltonian path: *A–B–C–D–E–F–G*

Other answers are possible.

Hamiltonian cycle: *A–B–C–G–D–E–F–A*

Other answers are possible.

ii Hamiltonian path: *D–E–C–B–A–F*

Other answers are possible.

Hamiltonian cycle: does not exist

iii Hamiltonian path: *A–B–C–D*

Other answers are possible.

Hamiltonian cycle: *A–B–C–D–A*

Other answers are possible.

MATCHED EXAMPLE 12

$$
\begin{array}{c c}
 & \begin{array}{cccc} P & Q & R & S \end{array} \\
\begin{array}{c} P \\ Q \\ R \\ S \end{array} &
\left[\begin{array}{cccc}
0 & 1 & 0 & 1 \\
1 & 1 & 1 & 1 \\
0 & 1 & 0 & 2 \\
1 & 1 & 2 & 0
\end{array} \right]
\end{array}
$$

MATCHED EXAMPLE 13

a C is connected to A, B and D, so every vertex is connected to at least one other vertex.

b 18

c The graph has an Eulerian trail because it has exactly 2 vertices with odd degrees.

d 7

e The edges are $AB \times 2$, $AD \times 2$, BC, BD, DC, AC, CC.

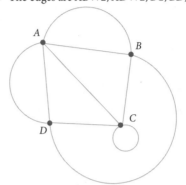

Other versions are possible.

MATCHED EXAMPLE 14

The shortest time needed for Peter to travel to Clara's house is 8 minutes.

MATCHED EXAMPLE 15

a The shortest time needed for Peter to travel to Clara's house is 8 minutes.

b
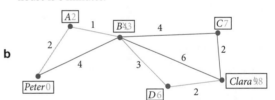

MATCHED EXAMPLE 16

a This is a spanning tree.

It has 8 vertices and 7 edges.

b This graph is not a spanning tree.

It has a cycle.

c This graph is not a spanning tree.

It has a multiple edge.

d This graph is a spanning tree.

It has 8 vertices and 7 edges.

e This graph is not a spanning tree.

It has an edge (*EG*) that isn't in the original graph.

f This graph is not a spanning tree.

It is not connected.

MATCHED EXAMPLE 17

A spanning tree has 4 vertices and 3 edges. The original graph has 4 edges, so we need to remove one edge to create a spanning tree.

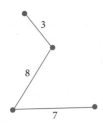

Total weight of spanning tree = 18

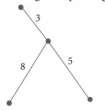

Total weight of spanning tree = 16

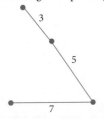

Total weight of spanning tree = 15

This isn't a connected graph, so it's not a tree.

The total weight of the minimum spanning tree is 15.

MATCHED EXAMPLE 18

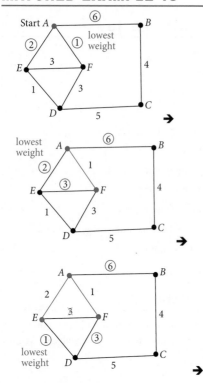

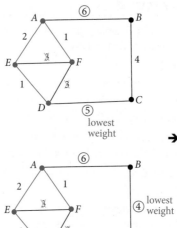

→

With the next step, all the vertices have been included:

This gives the minimum spanning tree as follows:

The total weight of the minimum spanning tree is $1 + 2 + 1 + 5 + 4 = 13$.

CHAPTER 10

MATCHED EXAMPLE 1

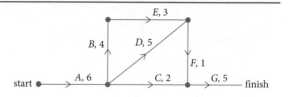

MATCHED EXAMPLE 2

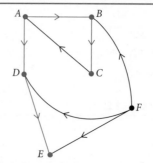

Vertices B, C, D and E are reachable from vertex A.

Vertex F is **not** reachable from vertex A.

9780170464055

MATCHED EXAMPLE 3

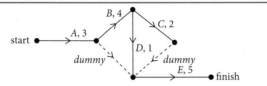

MATCHED EXAMPLE 4

To calculate the EST for the finish, find the total for the paths containing activity F and the path containing activity G. The highest figure obtained is the final EST.

Activity F: $12 + 2 = 14$

Activity G: $10 + 6 = 16$

Therefore, 16 hours is the final EST.

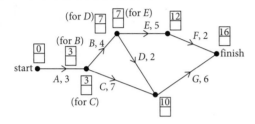

MATCHED EXAMPLE 5

Calculate the LST for activities B and C.

Activity B: $8 - 4 = 4$

Activity C: $10 - 7 = 3$

If there are two LST values at the end of an activity, then the smallest value is used to calculate the LST.

Therefore, use 3 hours to calculate the LST for the start.

The start LST $= 3 - 3 = 0$.

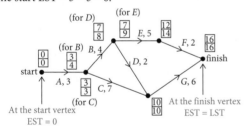

MATCHED EXAMPLE 6

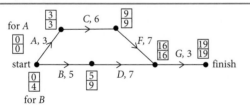

The critical path is A–C–F–G.

MATCHED EXAMPLE 7

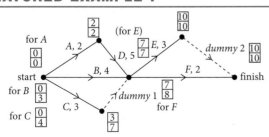

The critical path is A–D–E.

Float for activity $B = 3 - 0 = 3$ days

Float for activity $C = 4 - 0 = 4$ days

Float for activity $F = 8 - 7 = 1$ days

MATCHED EXAMPLE 8

a Critical path $= A$–D–F–G

Project completion time $= 27$ days

b The critical path is A–B–E–G and the project completion time is 18 days.

MATCHED EXAMPLE 9

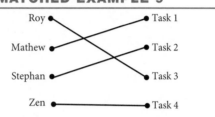

Task allocation:

Roy – task 3   Mathew – task 1

Stephan – task 2   Zen – task 4

MATCHED EXAMPLE 10

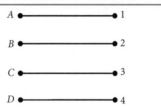

The allocations are $A1$, $B2$, $C3$ and $D4$.

MATCHED EXAMPLE 11

Z needs to complete task 2, but X and Y can complete either task 1 or 3. So, there are two possible allocations:

$X1$, $Y3$, $Z2$ and $X3$, $Y1$, $Z2$

MATCHED EXAMPLE 12

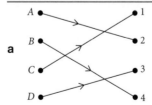

a

Allocation

$A2$, $B4$, $C1$, $D3$

b Total time $= 6 + 13 + 20 + 14$

$= 53$ hours

MATCHED EXAMPLE 13

A The total inflow capacity of vertex A

= 240+ 180 + 200

= 620 litres per minute

B The total outflow capacity of vertex A

= 250 + 270

= 520 litres per minute

C The flow out of vertex A = 520 litres per minute.

MATCHED EXAMPLE 14

Flow of cut

= 2 + 4 + 5

= 11

MATCHED EXAMPLE 15

The maximum flow from source to sink = 46.

MATCHED EXAMPLE 16

The shortest time path of 10 minutes is $A-D-E-Z$.